Agriculture and Food in the 21st Century

Agriculture and Food in the 21st Century

Economic, Environmental and Social Challenges

Festschrift on the Occasion of
Prof. Dr. Dr. h.c. P. Michael Schmitz 65th Birthday

Edited by Monika Hartmann and Joachim W. Hesse

Bibliographic Information published by the Deutsche Nationalbibliothek
The Deutsche Nationalbibliothek lists this publication in the Deutsche Nationalbibliografie; detailed bibliographic data is available in the internet at http://dnb.d-nb.de.

ISBN 978-3-631-64771-4 (Print)
E-ISBN 978-3-653-03559-9 (E-Book)
DOI 10.3726/ 978-3-653-03559-9
© Peter Lang GmbH
Internationaler Verlag der Wissenschaften
Frankfurt am Main 2014
All rights reserved.
PL Academic Research is an Imprint of Peter Lang GmbH.

Peter Lang – Frankfurt am Main · Bern · Bruxelles · New York · Oxford · Warszawa · Wien

All parts of this publication are protected by copyright. Any utilisation outside the strict limits of the copyright law, without the permission of the publisher, is forbidden and liable to prosecution. This applies in particular to reproductions, translations, microfilming, and storage and processing in electronic retrieval systems.

www.peterlang.com

Preface of the Editors

It is a great honour and pleasure for us to present this Festschrift at the occasion of Dr. Dr. h.c. P. Michael SCHMITZ's 65th birthday. This book is meant as a mark of gratitude for and recognition of his productive work over the last 40 years. With their articles his doctoral supervisor, academic students and their students as well as colleagues thank P. Michael SCHMITZ, who has been an inspiring research-partner, scientific advisor, and friend to them.

P. Michael SCHMITZ's professional career started 1975 when he concluded his diploma degree in economics (Diplom-Volkswirt) at the University of Göttingen and began the work on his doctoral degree which he earned at the same university in 1979 under the supervision of Ulrich KOESTER. In 1984 he was awarded his Habilitation at the University of Kiel. Professorships at the Institute of Agriculture Policy and Market Research, University Giessen and at the Institute of Agriculture Economics, University Frankfurt/Main followed as well as extended visits as guest researcher at the International Institute for Applied Systems Analysis in Laxenburg, Austria and at the Department of Agricultural and Applied Economics at the University of Minnesota (USA).

His research interest in the development of countries in transition motivated him to become the coordinator of the university partnerships Giessen with Kazan (Russia) and with Bila Tserkva (Ukraine). It has been P. Michael SCHMITZ's particular belief in the benefits of integrating business, research and education that has motivated him to found in 1995 the Society for Agribusiness-Research with the associated Institute for Agribusiness. The aim of this initiative has been to contribute to the strengthening of competitiveness of agribusiness by research and education. P. Michael SCHMITZ has been a long-time member of the Scientific Council of the German Federal Ministry of Food, Agriculture and Consumer Protection (1992-2012). During this time he contributed to numerous policy recommendations forwarded by this body.

At present P. Michael SCHMITZ holds the positions as Professor for Agricultural and Development Policy at the Justus-Liebig-University of Giessen, Director of the Center for International Development and Environmental Research at the same university, Director of the Institute for Agribusiness in Giessen and Director of the Institute of European Integration at the National Agrarian University of Bila Tserkva (Ukraine).

Those having had the privilege to work with him appreciate his clear economic thinking, his leadership as well as his human approach. His mission has been to train and develop researchers with their own mind. His students are grateful for the guidance and trust they received by him which provided the basis for their later employments in research, business and administration all over the world. Long before there were any special mentoring and training programs for female researchers he took a lead in the German agricultural economics profession in offering female students to work under his supervision on their doctoral degree, introducing them to his research networks and supporting them in their further careers. Thus, it is not surprising that two of his female students are among the first women who received a position as agricultural economics professors in Germany.

The papers in this book are from former students, six of them are themselves professors by now (and their students) and colleagues who have shared the same research interests as P. Michael SCHMITZ at various points in time. To the extent that several authors contributed to a paper in this volume it is always the first one having a special relationship with P. Michael Schmitz. The book covers theoretical as well as empirical work with respect to challenges to the food and agricultural sector, considering conflicts between social and ecological requests of citizens and market outcome, limitations and requests with respect to policy to tackle those conflicts and empirical evidence for several of those challenges that can contribute to a more evidence based policy formation. Those topics link nicely to the overall title of the book 'Agriculture and Food in the 21st century: Economic, Environmental and Social Challenges'. The contributions have not been professionally reviewed and editing of the papers has been limited to matters of format. The main body of the book is organized according to the two areas 'Agri-food markets and policies' and 'Agriculture, Trade and Development', each covering five papers. At the end of the Festschrift the reader will find the names of all contributors to this book and their affiliations as well as a list of publications by P. Michael SCHMITZ.

The first part of the book is devoted to a discussion about the complex interdependencies between agri-food markets and policies and thus to a research area which has played a central role in the work of P. Michael SCHMITZ over the last decades. Ulrich KOESTER, the doctoral supervisor of P. Michael SCHMITZ starts out with a contribution on 'Morals, Markets and Policies: Views with a Focus on Food and Agricultural Markets'. In his paper KOESTER discusses the relationship between morality and the outcome of market forces at an abstract level as well as considering the special conditions prevailing in the agricultural and food markets. The author shows that market outcomes might lead to consequences which are not in line with a given society's moral code. However, the limits of policy in this respect are also revealed. The author points to the necessity of an institutional framework that provides incentives compatible to the moral norms of society. Only under these circumstances can a 'happy marriage' between morals, markets and policies be expected in the food and agricultural sector.

The second paper by Monika HARTMANN, Johannes SIMONS and Kalkuli DUTTA titled 'Farm Animal Welfare: A challenge for markets and policy' investigates the conflict between market outcome and moral norms of society in the case of Farm Animal Welfare (FAW). The authors explain that governmental interventions for improving FAW can be justified if asymmetric information prevails, externalities exist and, to some extent, it shows non-private good characteristics. They also point to the limits of governmental interventions in a global world where regional policy interventions are constrained by WTO trade commitments. Thus, innovative solutions are needed to close the gap between reflective preferences or morals on the one hand and the market outcome on the other. Self-regulating strategies can help overcome the consumer - citizen gap by rising private standards for animal welfare if it is coordinated as shown using an example from the German meat sector. The paper reveals that the issues linked to FAW are far from trivial and their improvement need the effort of all members of the meat value chain as well as the government.

'From Policy Analysis to Recommendations for Evidence based Food Policy: Some Thoughts on „New" Policy Instruments' is the title of the contribution by Roland HERRMANN, Rebecca SCHRÖCK and Matthias STAUDIGEL. Using regulations for geographically differentiated foods and of food taxes as an example the authors investigate to which extent the conditions for an evidence-based policy are fulfilled. Evidence based refers to the effectiveness and efficiency of a policy. The authors show that theoretical considerations and empirical evidence strongly suggest that taxes on foods are neither an effective nor an efficient health-policy instrument. With respect to the protection of geographically differentiated foods they conclude that present EU policies are not evidence based as there is a lack of empirical studies providing insights into e.g. the income and welfare effects of regulations for geographically differentiated foods such as Protected Designation of Origin and Protected Geographical Indication

That policy is not always evidence based is also revealed in the paper by Michaela KUHL on 'Regulation of agricultural derivatives markets'. The author shows that significant pressure especially by non-governmental organizations on political and private actors and not scientific evidence has initiated new regulations on trade in derivatives and also led to decisions of many banks to withdraw from the market for speculative agricultural derivatives. Though the legal process is still inconsistent between countries and not yet completed KUHL raises the question whether these measures tackle the real problem of price instability and hunger or whether it might even aggravate them.

Christian FISCHER examines in his essay 'The future food value chain' the food supply and value chain concepts in the context of the global agricultural and food system. According to the author today's businesses and value chains are confronted with a fast changing environment. This has already led to a considerable heterogeneity regarding the types of businesses and value chains, a process which likely will continue. Also the way firms collaborate is changing. While traditionally, this was primarily referred to the question of "make" (yourself) or "buy" (from others), now the concept is to "ally" – i.e., to make together. FISCHER also discusses the change in meaning of the word 'value creation'. While this used to refer primarily to profit generation, there is an increasing recognition that capturing value includes the consideration of the social and environmental impact of a firm's activities.

The second part of the Festschrift focus on the link between agriculture, trade and development and thus as well at areas P. Michael SCHMITZ has considerably contributed to with his research. In their paper on 'Impacts of an EU-USA-Free Trade Agreement on Developing Countries' Martina BROCKMEIER, Tanja ENGELBERT and Janine PELIKAN investigate which economic effects might be created by the Transatlantic Trade and Investment Partnership (TTIP) currently under negotiation between the EU and the USA. The analyses are carried out using the static CGE model GTAP as a simulation tool. The authors direct their attention to two aspects; the impacts of such an agreement on developing countries and the importance of non-tariff trade barriers (NTBs) in the total liberalisation effects. Since adequate data for depicting NTBs in a CGE model are not available the ad valorem tariff equivalents of non-tariff trade barriers were econometrically estimated by employing the standard procedure of gravity modelling. The free trade scenario features the elimination of tariffs and a 50% cut of the NTBs between the EU and the USA. While these two trading partners show considerable gains in welfare and GDP due to TTIP the developing countries are not able to realise such improvements. The results clearly indicate that it is important to include NTBs in the liberalisation process.

Effects of a free trade agreement are also analysed in a second contribution. However, they are not the only ones Jong-Hwan KO focuses on in his essay 'South Korea Free Trade Agreement and its Impacts on Agriculture in Consideration of a Different Level of Regional Aggregation: A Computable General Equilibrium Approach'. In addition, he also pays attention to problems of spatial aggregation and assesses for this purpose the free trade agreement between South Korea and the EU which entered into force on July 1st in 2011. In his analysis KO represents the EU in 8 spatially different ways in the static CGE model GTAP and compares their free trade effects for Korea and for Germany. The aggregation levels used span from depicting the EU as one entity, in which all its member states are put together, up to the case where each member state is individually represented in the model. His major findings are indicative of the fact that aggregation matters. KO suggests using the most disaggregated data of the EU for such an analysis.

With his paper 'Food Security and WTO Domestic Support Disciplines post-Bali' Alan MATTHEWS picks up a long-term controversy regarding the question whether WTO rules and disciplines are consistent with the policy environment needed in developing countries to pursue their food security objectives. Though all three pillars of the Agreement on Agriculture are of relevance in this debate the author focuses on the domestic support pillar and thus, the issue addressed in the Bali Ministerial Conference. The author reviews the various proposals made for adjusting WTO rules regarding the treatment of procurements at administered prices undertaken to increase public food security stocks. The intention of those proposals is to increase developing countries' flexibility (policy space) in pursuing currently non-exempt policies if justified for food security purposes. There are two general options to widen a country's policy space in the domestic pillar; first by enlarging the scope of exempt policies and second by increasing the limits on its Aggregate Measure of Support (AMS) support. The message of the paper is that the former is the more preferable approach.

Khandaker M. M. RAHMAN, Mohammad I. A. MIA, Mohammad Z. ABEDIN and Mohammad Z. RAHMAN analyse Bangladesh's rice production systems from several points of view in their paper on 'Production and Wastage of Rice in Bangladesh'. To these different types of investigations belong an efficiency analysis of rice production and an assessment of losses at the different stages along the entire supply chain of rice. Finally, based on the outcomes of these analyses the authors formulate policy recommendations for improving Bangladesh's food security. According to the variation in the tasks different methods are employed. Most of the approaches are descriptive in nature. The efficiency analysis is based on a stochastic production frontier of the Cobb-Douglas type. Total loss of rice is rather high in Bangladesh. It amounts to slightly more than 30% of gross production. However, this share includes also - a so called - potential loss; i.e. lack of production due to inefficiencies. The causes of these losses are manifold and so are the policy recommendations forwarded by RAHMAN et al.

In his chapter on 'Climate Change Impacts on Agriculture and the Relevance of Adaptation: The Case of Pakistan' Mirza Nomman AHMED investigates the future impacts of climate change on Pakistan's agriculture and assesses past adaptation strategies. The quantitative results are obtained by employing the Ricardian climate change valuation technique. The study places special efforts on including all climate adaption strategies available to Pakistan's farmers and on providing variables for determining spatial effects across the country's Agro-Ecological Zones, provinces and farm-types using a multi-seasonal approach. Climate change scenarios are based on runs of a Global Circulation Model. They look far into the future, up to 2090. Results of the analysis provide detailed information for the rather diverse farming conditions prevailing across this country but also the different growing seasons. Increases in temperature during the winter growing season affect farming positively. Such a change has negative impacts if it happens during summer months. Also, rain-fed farming is considerably more vulnerable to climate change compared to the one using irrigation. AHMED concludes that farmers in Pakistan have well adapted to the current climatic conditions.

It goes without saying that we are grateful to all contributors to this Festschrift. Without their special commitment this book would not have been possible. Likewise we thank Alexandra BENDER for taking over the formatting activities and Palina MOLEVA for her editorial work. The tangible support provided by the Society for Agribusiness Research (Verein für Agribusiness-Forschung e.V.) is also gratefully acknowledged.

Monika HARTMANN and Joachim W. HESSE

Table of contents

Preface of the Editors 5

Agri-food Markets and Policies

Morals, Markets and Policies: Views with a Focus on Food and Agricultural Markets 13
Ulrich KOESTER

Farm Animal Welfare: A challenge for Markets and Policy 37
Monika HARTMANN, Johannes SIMONS and Kakuli DUTTA

From Policy Analysis to Evidence-based Food Policy Recommendations: Some Thoughts on 'New' Policy Instruments 61
Roland HERRMANN, Rebecca SCHRÖCK and Matthias STAUDIGEL

Regulation of Agricultural Derivatives Markets 89
Michaela KUHL

The Future Food Value Chain 101
Christian FISCHER

Agriculture, Trade and Development

Impacts of an EU-USA-Free Trade Agreement on Developing Countries 107
Martina BROCKMEIER, Tanja ENGELBERT and Janine PELIKAN

EU-South Korea Free Trade Agreement and Its Impacts on Agriculture in Consideration of a Different Level of Regional Aggregation: A Computable General Equilibrium Approach 129
Jong-Hwan KO

Food Security and WTO Domestic Support Disciplines post-Bali 163
Alan MATTHEWS

Production and Wastage of Rice in Bangladesh 185
Khandaker M. M. RAHMAN, Mohammad I. A. MIA Mohammad Z. ABEDIN and Mohammad Z. RAHMAN

Climate Change Impacts on Agriculture and the Relevance of Adaptation: The Case of Pakistan 209
Mirza Nomman AHMED

Contributors 243

Bibliography of P. Michael SCHMITZ 245

Morals, Markets and Policies: Views with a Focus on Food and Agricultural Markets

Ulrich KOESTER

Abstract

Morality and markets are widely held to be two unrelated concepts or, if related, then their interaction is considered more negative than positive. This perception has contributed to the bad esteem of economists. This long-term tradition of neoclassical economics – which is still the mainstream in economics – posits that economic agents, although they may be highly selfish, nevertheless offer society significant benefits. Agricultural markets' functionality is often in the centre of criticism and food scandals seem to support the view that economic agents do not serve the common interest.

The goal of this article is to clarify the relationship between morality and the outcome of market forces under specific conditions on the one hand and real world conditions on the other. It emphasizes that a clear definition of morality is essential to the accurate assessment of market functions from a moral point of view. Moreover, a positive relationship will prevail only under specific market functioning conditions which do not exist in most markets, and are particularly absent in agricultural markets. Based on these findings, we explain why agricultural markets are exceptional and not always in line with commonly held moral understanding.

Because agricultural markets require specific governmental interventions to provide for incentive compatibility of private agents, the government must create an environment in which selfish actions are considered acceptable from a society's point of view. The second last section of the paper proposes a design for EU agricultural policy which might perform in harmony with the morality of a given society. It argues that policymakers who behave like private entrepreneurs must be forced by a specific institutional framework to make decisions that are best for the public at large. In the final section, we discuss selected EU agricultural policy changes introduced for fiscal years 2014 through 2020 from a morality point of view.

Acknowledgements

The author thanks M. HARTMANN, Bonn, W. BRANDES, Göttingen, B. SCHULZE, Kiel and A. PELLILLO, Tbilisi for valuable comments on an earlier draft of the paper. The article gained significantly from many fruitful discussions with R. MÜLLER, Kiel.

Introduction

During the last decade, public trust in markets in general – and in financial and food markets in particular – seems to have deteriorated. Food scandals feed mistrust and strengthen the commonly held belief that markets support immoral behaviour, leading to an increased call for governmental intervention despite a widespread mistrust in politicians. Strangely enough, not long ago, at the time of the collapse of the socialist economies, the benefits of a market economy were praised. The reputation of economists has also deteriorated as mainstream economists continue to stick to their models, assuming that rational decision makers react to market prices while simultaneously contributing to the economy's overall welfare. This article will clarify relationships between morality and the outcome of market forces under specific conditions on the one hand and under real world conditions on the other. Based on these findings, we will explain why agricultural markets are exceptional and not always in line with commonly held moral understandings. Its second last section is devoted to the design of an agricultural policy which might perform in harmony with the morality of a given society. It argues that policymakers who behave like private entrepreneurs must be forced by a specific institutional framework to make decisions that are best for the public at large. The final section discusses selected EU agricultural policy changes introduced for fiscal years 2014 through 2020. A summary will close the article. It has to be emphasised that the reasoning focuses mainly on private goods; the problem of public goods is only taken up in the last sections which deal with policies.

Definition of Morality

Unfortunately, no generally accepted and controversy free definition of morality is available. However, a clear definition of morality is essential to any investigation of the relationship between morality and markets. Therefore, it is quite clear that this definition must be based on personal judgment which will, of necessity include value judgments.

It is agreed that morality is a reflection of a given society's norms which expresses its values based on a system of rules and moral guidelines broadly accepted by that society. Therefore, morality affects and may constrain people's social behaviour (GRIFFITHS et al., 2001 and BARON, 1998).

Morals serve as the rules of the game which, whether formally laid down in legislation or evolved over time in the form of unwritten tradition. Based on acceptance of this definition, there is no "bad" or "good" morality; rather morality is merely a description of a society's accepted norms (GERT, 2012). It is quite clear, then, that these norms differ across societies and will change over time as they are partly determined by culture.

Individual behaviours can be classified as morally "bad" or "good" depending on the norms of the society under consideration. Whether a person's behaviour is considered morally acceptable or immoral depends on the judgment of the society of which that person is a part. As norms differ across societies and over time, a specific behaviour may be judged morally good by one society, yet bad by another. Furthermore, a given society may view the same action differently at different times, depending on how that specific society is defined. For example, it might be merely a group of people who share certain values which are not necessarily identical with those of the population in a specific country. The rural population may adhere to different values than those of the urban population. People belonging to different religious groups may favour still different values. In general, one person's act can be morally good or bad only if other people are affected.

Even in cases of general agreement on moral standards within a given society, specific criteria which can be used to judge whether an individual has acted in a bad or good way from a moral point of view must be defined. In defining morality, philosophers seem to agree on two different and important elements by which morality may impact an individual's action:

- A person may be motivated to do good things in line with moral standards of a given society and may select, from among the possible alternatives, those which are compatible with his motive.
- Moral people behave according to the expected consequences which are in line with the moral codes of the society where they live.

According to the school of deontology, motives are crucial for judging moral behaviour. Kant, one of the scholars of this school, postulates that an individual's motives should act as a personal framework for her/his actions even if those consequences are not appreciated by other members of society. Of course, this definition includes a strong value judgment which may not be shared by many people. Hence, this criterion is somewhat modified by the school of moral relativism which, in principle, agrees on the dominance of motives, but also takes into account the action's consequences. Finally, in contrast to the other schools, consequentialism (or teleological ethics) focuses only on an action's consequences. Whether the individual actor must intend for others to judge her/his action as morally positive or whether only the consequences matter, regardless of his motive, may be open to discussion.

Economists may argue that today's world is too complicated and, therefore, the individual cannot always anticipate consequences. Well-intended actions by do-gooders may lead to bad consequences, while very selfish actions may result in outcomes which are highly appreciated by society. As early as 1714, Mandeville was one of the first authors to emphasise that private vices form the basis of overall welfare (quoted by SEDLACEK, 2012). In fact, do-gooders may actually place hardship on the backs of others; hence, we must accept that well-intended actions may lead either to socially desired results or to undesired ones. In contrast, however, highly selfish actions may also lead to either good or bad results.

Economists tend to be more practically oriented, and mainstream economists trend to emphasize the consequences of individual actions independently from the motives of the actor and the intended consequences. It does not matter why an individual acts in a specific way. Whether it may be self-interest or altruism, only the outcome is important. Whether the action is considered good or bad depends on how other members of the society value the outcome. If they rate the outcome as acceptable from their point of view, the individual's actions are considered morally good.

It should be noted here that this definition of morality does not imply that any activity which leads to Pareto improvement – as defined by economists – is morally good. Pareto improvement is implicitly based on socially independent individual utility functions which include only material welfare. However, it might well be that although an individual's income is not affected by other people's actions, that individual may, nevertheless, disapprove of the specific actions of others if distribution has been affected. Hence, the definition of morality just put forward does not imply that individual welfare is socially independent. People may even be concerned about actions of others, even if their material welfare is not affected, because they consider the enrichment of specific people as unfair.

If we accept – in accordance with mainstream economics – that only the consequences of actions, and not their motivation, are important, it may well be that an individual who acts only in his own self-interest may inadvertently act in the interest of others. However, this coincidence will occur only in cases of incentive compatibility, meaning that incentives for individuals should lead to actions and consequences which are in line with the common interest. Hence, in the following, we must evaluate the conditions under which markets lead to incentive compatibility, whether these conditions hold true in the real world of markets in general, and of agricultural markets in particular.

Conflicts between morally-constrained and market-oriented behaviours may disappear, either partly or completely, if a) one agrees with the definition of morality as put forward by economists, and b) if it can be shown that the actions of those who serve only their own self-interest and react to market incentives may lead to positive consequences, even from a moral point of view.

This leads us to the first interim conclusion: Whether markets and morals can engage in a happy marriage depends, first of all, on the definition of morality. The consequentialist position must be accepted. Moreover, it must be assumed that whether the consequences were intended or unintended is irrelevant.

However, even if the economists' definition of morality is accepted, it cannot be taken for granted that, from a morality point of view, markets can always be rated positively. This problem will be investigated in subsequent sections. But first, the relationship between morality and the effects of markets will be analysed.

The relationship between morality and the functioning of markets in general

Mainstream economists like to refer to the father of modern economics, Adam Smith (SMITH, 1776), whose well-known example posited that society may even be better off if individuals act selfishly. However, he did not say that, under all market conditions, selfish actions would always lead to consequences which were appreciated by the society at large (see in particular SMITH, 1759). Modern economics have proved that, under certain conditions, selfish actions guided by market incentives will lead to a social optimum where nobody's position can be improved without negatively influencing the welfare of others (DEBREU, 1952). It should be noted that the findings of so-called neoclassical economics are based on specific assumptions used in mathematical models. It is widely accepted that markets in the real world may not lead to the desired outcomes.

Neoliberal economists have defined certain tasks for the state so as to establish an enabling environment for possibly acceptable market outcomes from the point of view of the society at large. This school of economics accepts the possibility that the market may fail and, hence, that there is a role for the government in reducing the impact of such failure. However, the danger of policy failure is also highlighted.

The neoclassical school focuses implicitly on one specific element of morality – individual freedom – including selection of the best alternative among many choices. It is probably true that most people value individual freedom very highly, but some individuals may prefer a degree of paternalism in exchange for more job security, more equal distribution of income and less pressure to change. If one accepts the stated axiom, that individual actions are morally good only if they are good from the society's point of view; one cannot assume that selfish actions can be accepted as morally good per se.

The following section discusses the interdependence between morality in a society and the functioning of markets. Figure 1 shows the importance of morality for the functioning of markets and the importance of functioning markets for morality within a society.

Markets are used to coordinate individual actions, facilitate cooperation and generate productive interactions. Hence, functioning markets have the same task as does the morality of people: both constrain individual behaviour and make it predictable. It is assumed that trust is the main element of morality which is of importance for the functioning of markets. Individuals may be trustworthy if they are honest, reliable, and punctual. Individuals may trust in the stability of the political and economic system if they trust policy makers.

Figure 1 shows that morality contributes to the reduction of transaction costs, giving rise to the evolution of markets. Exchange of goods is generally based, either explicitly or implicitly, on a contract, even if not in a written form. Partners in an exchange trust in the honesty of the partner with respect to the quality of the product offered and in the behaviour of the partner of exchange. It is expected that the partner of exchange will fulfil his commitments completely and on time. This point is of special concern if the exchange involves agricultural product and factor markets.

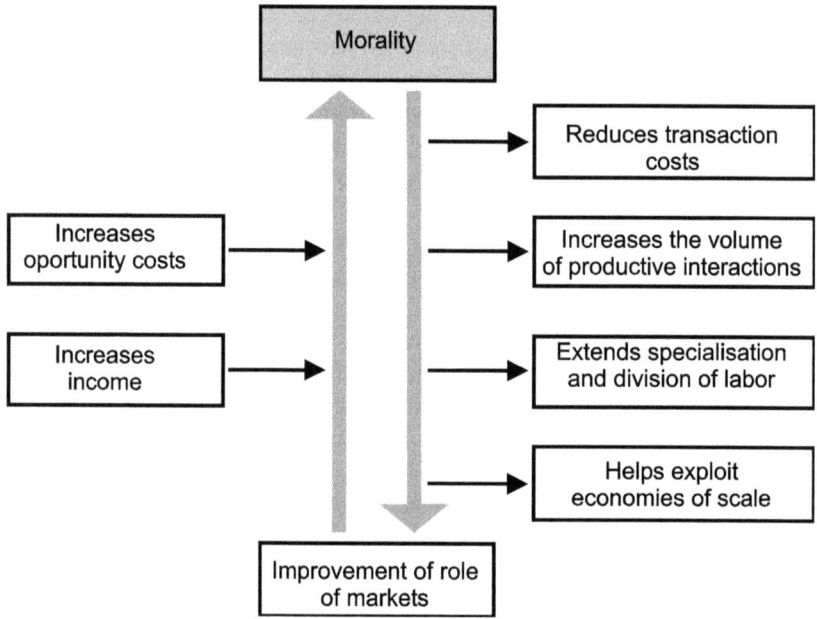

Fig. 1:   *Interdependence between morality and markets*
Source: KOESTER, U. (2014)

Figure 1 also indicates that markets may contribute to added morality in a society. For example, if an individual is involved in numerous transactions, he has opportunities to build up his reputation and trustworthiness. Contracting partners may spread information about his behaviour among potential partners of exchange, or market information systems may inform potential contracting partners about the quality of the offered product as well as the trustworthiness of the seller.

This same scenario increases an individual seller's consequences if he is untrustworthy. News about the immoral behaviour quickly spread among potential buyers, and the seller loses business. Ergo, the risk of lower profit may enforce trustworthiness.

If the moral infrastructure is not available, it is likely that significant problems will hinder the development of markets (FRIEDMAN, 2008). Transition to a market economy was generally very negatively affected by the lack of moral infrastructure. Privatisation, which is generally considered a prerequisite for creation of a market economy led to a "transition to kleptocracy" as described by Friedman (FRIEDMAN, 2008). This observation hints at a dilemma: A certain standard of morality is needed if markets are to be set up and if the outcome is to be acceptable from the society's point of view. However, the necessary standard of morality might not be available because markets to enforce morality did not exist.

So far, the analysis shows that morality and markets are not necessarily in conflict; rather, in many cases, they are complements. Markets are partly driven by morality and markets may enforce morality.

Specifics of Agricultural Markets and Morality

Most economics textbooks deal exclusively with search goods, meaning the buyer knows the quality of the product and only needs to search for the lowest price. In contrast, most agricultural products are experience or credence goods. Expected quality of experience goods may be ascertained by the buyer only after some time has elapsed. In the case of credence goods, the buyer must rely on information with respect to the not visible production and processing method and the ingredients of the products. The latter play an increasingly important role in agricultural and food markets. This is especially apparent for food which is labeled as organically produced. The consumer has little chance of discovering whether the seller's promises are trustworthy. In case, for example, a product has been declared organically produced, the buyer cannot determine easily – if at all – whether the production process was actually as declared. He must trust in the seller's declaration. These markets may not evolve without trust, and transactions will be limited. This explains why countries which may not use any modern production methods, such as mineral fertilisers or chemicals due to their low stage of development and thus underdeveloped institutions are unable to sell their products as organically produced.

From the consumers' point of view, the food market situation has become more complex. Today, many consumers are more health-conscious than they were many years ago, at times with lower income. The food industry seems to support consumers by specifying the quality criteria on the label. Some of these criteria, such as bio-label, indicate that the product must have specific attributes which help consumers make their choices. Quality is ensured largely by detailed specifics and legal controls; however, many labels may misinform the consumer, as various studies and tests reveal[1].

WISO (2014), a TV program, reported on a test undertaken by WISO and the German Gesellschaft für Konsumforschung. There are some food producers which legally misinform consumers[2]. Labels stating, for example, "from controlled production", "recommended by customers," or "100 percent tested quality," suggest that certain products are of superior quality compared to those without such label. In fact, however, there is no proof of any superior quality of these products. According to EU legislation, from December 13, 2016 onwards, voluntarily-provided information must meet the following requirements:

- It shall not mislead the consumer.
- It shall not be ambiguous or misleading.
- It shall, where appropriate, be based on relevant scientific data.

Until that time, consumers can be deceived by specific labels[3] (EUROPEAN UNION, 2013). The relevance of misleading labels on the European food markets is revealed by experiences of the European Food Safety Authority (FSA). Food producers who use labels with health claims must obtain approval from FSA. According to Food Watch, FSA has, so far, tested 2,500 food products which had been advertised with health claims, but only about 250 have been approved. Based on this evidence, 90 percent of food products tested to date deceive the consumer regarding health claims. This experience clearly supports the idea that many food products are still credence goods and that governmental control can improve trust in food products.

It should be noted that changes in favour of consumers can hardly be expected to occur via changing the attitude of food processors through appeals. Although appeals may have some effect, a wealth of evidence demonstrates that individual behaviour is unlikely to respond to requests as long as the consequences of immoral but legal actions are profitable. Measures that increase opportunity costs

---

1 See the study by ZÜHLSDORF et al. (2013): Ergebnisbericht. Kennzeichnung und Aufmachung von Lebensmitteln aus Sicht der Verbraucher – Empirische Untersuchungsbefunde.
2 The German Ministry for Food and Agriculture has set up a website which informs consumers of specifics of labels on food products, so does the website by Verbraucherverbände. A study by the company SGS (2014) informs on the main concerns of consumers.
3 See also the review by GILSEMAN, 2011. Nutrition & health claims in the European Union: A regulatory overview.

for immoral behaviour are needed. This holds true, not just for legal, but also for immoral and illegal behaviour. Laws and their enforcement are needed.

Food scandals also prove that farmers or food processors may have used food product ingredients differently than as legally allowed and/or farmers may have used inputs which were not legally allowed at the time of production. Obviously, these markets suffer from incomplete information, leading to cases of unbalanced results. Concerning food safety and illegal behaviour, European countries may consider changing laws concerning whistle-blowers (WORTH, 2013). Illegal actions within the food market are often not carried out mainly by farmers who do not employ wage earners, but by enterprises with some or even many wage earners. If, for example, a food company offers lasagne which includes horse meat instead of beef – which was one of the main food scandals in 2013 - many workers must be aware of this criminal action. Reducing the uncertainty for whistle-blowers with laws which improve protections for these persons over those currently extant could help reduce illegal actions.

In addition, it is possible that some farm inputs used in the production of food may, at some time, not have been in line with the nations' health standards. The BSE (Bovine Spongiform Encephalopathy) scandal that took place in Europe in the late 1980s and early 1990s was one of the most outstanding cases. The use of mammalian protein in the manufacture of animal feeds given to ruminant animals, such as cows, sheep, and goats is now forbidden. Due to new research findings, the use of seed and chemical inputs in the production processes taking place on the farms is changing permanently. The changes are introduced because farmers expect a higher profit. However, food safety concerns may militate against the use of these inputs. Hence, testing is needed for new inputs, such as seed, chemicals and feed, as well as new foods and their ingredients. Strong control of agricultural production and food along the food supply chain is an important task of the government and will help reconciling morality with markets.

Additionally, trust is of high importance for changes in the structure and efficiency of agriculture (see AHMED et al., 2010). Prospering agriculture is based on increasing use of variable inputs, such as labour, land, capital, fertiliser and others. These agricultural input markets will evolve only with trust among potential trading partners. Experience in transitional countries has shown that restructuring agricultural sectors largely depended on reorganising the farms. However, as successors of the former collective farms, the newly founded farms did not have access to credit and were, therefore, unable to restructure. These farms had no opportunities to prove their creditworthiness, which is built mainly on trust, and they could not offer any collateral. There were, however, some exceptional cases: some managers of new farms who did not own the farm's assets and could not offer the farm's capital as collateral offered their private capital, in most cases privately-owned houses, as collateral. Some banks considered this offer a strong commitment to serve the debt and were willing to grant credits.

Lack of trust has also delayed farms' reorganisation in transition countries because transfer of land was very limited (SCHMITZ, 1998 and 2011). Land is highly property right intensive. If property rights are not well defined and secured, the evolution of a land market is suppressed (OLSEN, 2000). A recent study by Transparency International (2011) reported on the effect of corruption, both within a country and in functioning land markets. The level of corruption can be taken as a good proxy for lack of trust. The study revealed a high correlation between the level of corruption and the functioning of land markets in countries included in the study. Consequently, efficient use of land is strongly determined by the governance of land, and the latter is strongly affected by the country's extent of corruption.

Corruption and lack of trust have likely contributed to a significant amount of land lying idle in many transition countries. Unfortunately, no exact information on the amount of unused land in individual transition countries is readily available. However, estimates are available for some countries, such as Russia, Ukraine, and Kazakhstan. At Green Week in Berlin 2014, Deputy Minister Sen reported that unused land may amount to up to 28 million hectars, of which one third could reasonably be activated. VISSER and SPOOR (2010) provided estimates by FAO and World Bank, which stated that the increased amount of unused land in 2005, compared to the period of 1990 to 1992, was quoted as 20 million hectares. However, on BBC, 1st August 2008, Daniel Fischer reporting on only Russia, stated that 100 million acres of fertile agricultural land was lying fallow (= 40 million hectares) (VISSER and SPOOR, 2010). SYSOYEVA (2011) reported that 21% of Russia's arable land, 20.7 million hectares, is unused. Even with no accurate estimates, the evidence is clear: A significant amount of land is unused, and one may wonder why this is true. Why is the present owner of the user rights not interested in leasing out the land? Some hypotheses for Ukraine will be provided.

Because of the moratorium on land sales, the selling of land is not allowed. Lease prices are set officially by the government based on officially stated, but ineffective, land prices. One likely hypothesis is that the person who owns the user rights is not willing to lease out the land. He may compare the present lease price with potential future land and lease prices that will prevail when the moratorium ends and, if he is uncertain whether the lessee would either pay the higher lease price, or whether he would be able to end the lease contract at that currently unknown time; otherwise he might be willing to sign only short-term lease contracts. However, potential lessees might not be interested in a short-term contract at a given price. The gross margin of land which has been idle for some time might be marginal because the soil will need to be improved by fertilisation and possible irrigation, which may take years. Hence, a situation such as that depicted in Figure 2 may prevail.

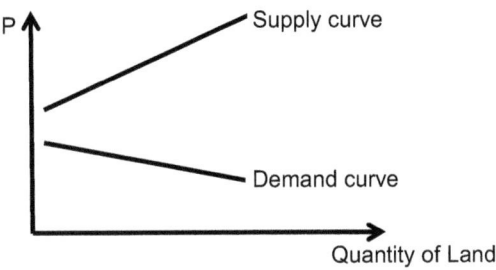

Fig. 2:   *Situation of the land market in Ukraine*
Source: Own presentation

The offer price which, among others, is determined by the expected lease price following the end of the moratorium, might be higher for all hypothetical prices than the demand price. Consequently, there will be no transactions on the land market, resulting in idle land in specific locations.

Lack of trust, combined with resulting high transaction costs on farms, have contributed to the survival of family farms in Western Europe and North America. Diversified and locally spread agricultural production has kept the cost of monitoring and enforcing labour contracts high, giving family labour a comparative advantage. Family farms preferred to forgo economy of scale benefits based on decreasing transformation costs due to high farm transaction costs. Of course, whether it is economically profitable for family farms to continue depends, not only on transaction costs for labour, but also on external transaction costs associated with buying and selling agricultural outputs and inputs, as well as the actual transformation costs. As transaction costs, both on and off the farms are strongly affected by trust and the legal system, it should not be surprising to see an evolving farm structure in most transition countries which differs markedly from that in Western market economies.

The state of the present analysis leads one to conclude that: Trust is most important for the evolution of markets and, in particular, of agricultural product and input markets. Agricultural and food products belong increasingly to the group of experience and credence goods. Governmental activity is needed for adequate functioning of these markets.

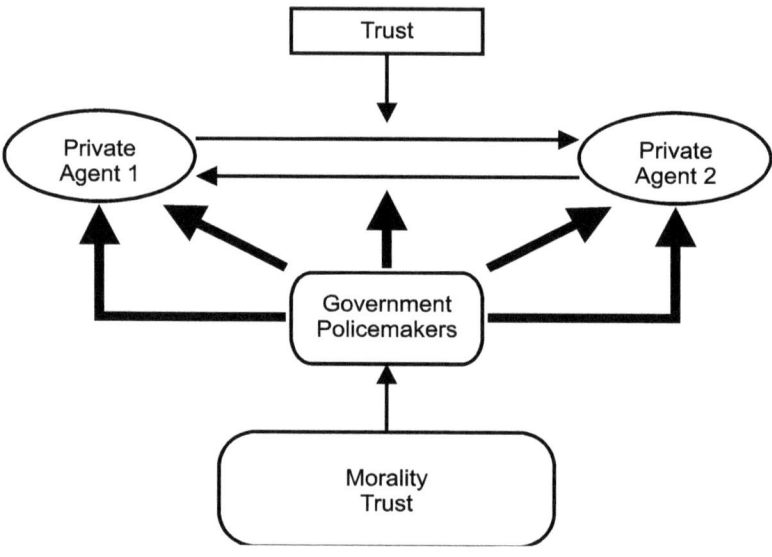

Fig. 3: *Relationship between trust, functioning of markets and the role of the government*

Source: Own presentation

The existence of this dilemma is important in assessing the relationship between morality and market performance and vice versa. There is little doubt that mutual enforcement depends largely on state activity. Whether government actions will actually contribute to a better marriage between markets and morality depends on two somewhat different issues. One concerns the design of the government. For example, what are the rules for political agents as part of a government and how are these rules enforced? The second concerns actual policy measures undertaken by a specific government to improve interaction among private agents. Of course, the rules for policy makers have an impact of the selection of policy measures, but the same rules could lead to the selection of different policy measures. Figure 3 illustrate the issue.

The design of the institutional framework for constraining political agents

If the government has to play a crucial role in a market economy, we must be assured that specific principles of a democratic society are followed. Good governance requires the rules of the game to be well defined for a well-functioning government. These rules could be based on:

- Transparency
- Accountability
- Avoidance of conflicts of interest in political decision making
- Evidence-based policy decisions

The importance of the institutional arrangement for constraining policy makers

Policy agents should act on behalf of the society at large. If it is accepted that agents must respond to the will of the population, the electorate is the principal of the political agent. The political agent will be most likely to act in the interest of the principal if incentive compatibility is ensured. The agents' personal interests should lead to actions which are acceptable for the society at large. Such a situation is most likely to prevail only if the principal is well informed about the agents' actions and their consequences. Hence, the principal wants to know what has been decided, how it has been implemented, and what the effects are. Whether the agent is likely to be guided by incentive compatibility depends largely on the institutional framework, which may support such actions.

Since the Common Agricultural Policy may serve as a good example, it may be of interest to investigate whether the institutional framework for specific decisions in the policy making process is adequate with respect to transparency, accountability of decision makers, and whether incentive compatibility for decision makers prevails.

Institutional constraints for the Council of Agricultural Ministers

The Council of Agricultural Ministers – the only legislative institution for agricultural matters up to implementation of the Treaty of Lisbon (December 1, 2009) – decided behind closed doors. Hence, the public was not informed as to which of the Ministers voted for specific decisions. Moreover, the public did not know on what arguments and empirical evidence the decisions had been based. This decision making structure may have allowed the Council to respond more to the Ministers' individual objectives than to the interests of the public at large. Actually, there is a strong hypothesis which posits that the objectives of the Agricultural

Ministers corresponded more with the interests of the farmers than with those of the society at large; the ministers' backgrounds may support that hypothesis. The German Ministers of Agriculture used to have strong agricultural backgrounds; however, over a period of 48 years, only two ministers did not have affiliations with agriculture – either by being a farmer themselves, by having been strong ties to the farming community.

Agricultural Ministers corresponded more with the interests of the farmers than with those of the society at large, and the ministers' backgrounds may support that hypothesis. The German Ministers of Agriculture used to have strong agricultural backgrounds; however, over a period of 48 years, only two ministers did not have affiliations with agriculture – either by being a farmer themselves or as offspring of a farm family. One of the exceptional ministers served for four years and the other for only 14 days. Hence, there is a strong presumption that the agricultural decision makers may have had a biased view in favour of the farm sector and, therefore, suffered from conflicts of interest. Decisions made behind closed doors may have enabled them to serve their clientele better than would have been possible in an environment with open doors that admitted journalists and the public. It was nearly impossible to hold individual ministers accountable for the consequences of their decisions because it was unknown how he/she had voted.

The electorate also suffered from lack of transparency concerning the effects of the policies. The principal normally wants to know why the agent has acted in a specific way, what consequences are expected, and what was realized. The CAP decision making procedure suffered strongly in this respect. Up to the MacSharry reform of 1992, changes in the CAP were dominated by achievement of the agricultural income objective. However, neither the Treaty of Rome nor the following Treaties mentioned the income objective as a first order objective. Instead, it was postulated in Article 39 (THE TREATY OF ROME, 1957) and the following treaties. "The objectives of the common agricultural policy shall be: a) to increase agricultural productivity by promoting technical progress and by ensuring the rational development of agricultural production and the optimum utilisation of the factors of production, in particular labour b) thus to ensure a fair standard of living for the agricultural community, in particular by increasing the individual earnings of persons engaged in agriculture." The article's phrasing states quite clearly that the first objective of the CAP should be to increase agricultural productivity and thus ensure a fair standard of living.

Up to the year 2014, the Council of Agricultural Ministers has never defined what a "fair standard of living" would be and how it could be quantified. There is hardly any disagreement that the standard of living is not reflected only in the average level of agricultural income; rather, it must take other variables into account, including income from off-farm activities and the value of the property. Moreover, information on averages is of no value for policy making if the variance across the population is large and policies are guided by social concerns.

It is well-known that the price support policy which was the main instrument employed by the CAP up to 1992, increased the variance of income across agricultural farms and levied poor households with a higher share of income spent on food than high income households.

It can be stated that, up to 1992, the CAP decisions were not transparent and not evidence-based, decision makers were not accountable and may have suffered from conflict of interest. Policy makers lacked adequate information about the actual situation, the diagnosis and the impact of the policies on the objectives. The institutional design was not adequate to support decisions in line with moral standards.

Price support and morality

The main measure employed by CAP was price support up to 1993. Hence, it will be explored whether this instrument is in line with securing incentive compatibility as a requisite for a happy marriage between markets and morality.

The impact of price support policy on incentive compatibility within an individual country

In line with neoclassical economics, we assume that producers try to maximise profit and consumers try to maximise utility. Producers confronted with competitive markets try to equate marginal costs with market prices, and consumers try to equate marginal willingness to pay with market prices. Individual decisions of these parties will only match with the interest of the society at large if market prices are equal to shadow prices of the overall economy. The shadow price expresses the marginal cost for a society to consume an additional unit of the product under consideration. Because the shadow price for a relatively small country is equal to the world market price, incentive compatibility requires that market prices be equal to world market prices.

Incentive compatibility, one of the necessary conditions for markets to be judged positively from a moral point of view, cannot prevail if price support is used to achieve distributional income effects. Consequently, price support leads to immoral decisions. The negative impact on morality is stronger if the nominal rates of protection across products differ largely, as in the CAP. EC decision makers were constrained by international agreements, namely the GATT (General Agreement of Tariffs and Trade), for using price support. The EU was not allowed to increase external protection for cereal substitutes. Increasing protection for cereals and those agricultural products which used inputs based on either cereal or cereal substitute increased the actual effective rate of protection for livestock products. The outcome was that producers received high incentives to produce these products which exceeded the demand of internal markets. The exportable surplus fetched prices on the world market which, in some periods, did not even cover the amount paid for the import of cereal substitutes to produce the surplus. It is obvious that this lack of incentive compatibility reduced domestic welfare.

Of course, producers cannot be blamed for producing this surplus because they merely responded to incentives set by policy makers. However, from a morality point of view, those who decided on these incentives are, indeed, to blame. Also from a distributional point of view price support can be questioned. This main instrument employed by the CAP up to 1992, increased the variance of income across agricultural farms and levied poor households with a higher share of income spent on food than high income households. The transfer of income, which may be acceptable on political grounds, could have been achieved with instruments incorporating higher transfer efficiency.

Interim conclusion: Price support violates the principle of incentive compatibility and can hardly be justified on the impact on distribution of income. Consequently decisions of farmers and food producers are not in line with the interest of the society at large.

Price support and divergence of interest among member states

If agricultural price support leads to an increase in the divergence of national interests, the external protection may conflict even more with incentive compatibility and morality standards. In 1977 Koester showed that a change in protection for an individual product may affect the distribution of income across the member states (KOESTER, 1977). An increase in the nominal protection rate for a specific agricultural product affects national income depending on the national and EU degree of self-sufficiency for that commodity, the elasticity of supply and demand in the country and in the EU, the actual degree of nominal protection and the national share in the common budget. In general, countries with high agricultural import dependencies and high shares in common finances will suffer more from an increase in external protection than will countries with high export orientation and a low share in common financing. Consequently, the chosen instrument of price support widened national interests among member states.

This explains why Great Britain, a general loser of price increases for agricultural products due to high import dependence, tried to close the need for imports by stimulating domestic agricultural production in the 1970s and 1980s. This kind of national policy was beneficial for Great Britain but not for the EU, as policy makers did not improve incentive compatibility, but changed it for the worse.

Concerning the evaluation of the CAP from a morality point of view one can conclude: The institutional framework and the selection of the main instrument, namely price support, contributed to incentive incompatibility between the interest of policymakers and that of society at large. The main outcome of policy decisions has to be judged negatively from a morality point of view. It seems that policymakers did not have the right incentives to create an institutional framework which led to consequences in line with moral standards.

The CAP has changed significantly over time and, due to lack of place, it is not possible to sequentially assess the individual changes. Therefore, this article will discuss only some of the most recent changes decided in 2013, with assessment based on the following criteria:

- Does the intervention improve incentive compatibility in the private sector and, thus, improve the functioning of markets with respect to allocative efficiency, including environmental effects? This criterion is together with the ones below based on the notion that it is the task of policy makers to create institutions which harmonize decisions of individual market agents with the interest of the society at large.
- Is it an evidence-based policy? This also concerns the capacity of the government to govern policies. This criterion implicitly claims that political agents are required to act in the interest of their principal, which is the electorate.
- Does the intervention change income distribution in line with the principles of a social market economy? The change would be positively assessed if the transfer efficiency of the income transfer were as near as possible to one. The justification of this criterion issues from the fact that income distribution plays an important role in the acceptance of a market economy and, hence, governmental intervention must take distributional effects into consideration.[4]
- Is the policy design adequate to allow for achievement of non-economic policy objectives with lowest economic costs? This last criterion concerns efficiency which seems to be at the heart of economists.

---

4 Some philosophers, A. SEN in particular, even suggest that governmental intervention should only be set in place if the income of the poorest segment of population in the society is positively affected (SEN, 2011).

- It must be emphasised that any evaluation of policies is normative. If the criteria are not accepted, the entire outcome of the evaluation may be rejected. However, those criteria are in line with moral behaviour as defined above. It may well be that specific policy decisions may not comply with all the criteria just mentioned as trade-offs may occur between compliance with individual criteria. However, those alternative policy decisions should be favoured that create fewer conflicts.

Selected major policy changes in the EU assessed from a morality point of view

The change in direct income support

To understand the adopted changes, a short history of direct payments in the CAP may be helpful. This budget item came into existence in 1993, as the result of the Council of Agricultural Ministers' 1992 decision to reduce the intervention prices for grain by about 33 per cent and to also reduce the support price for oilseeds. In 1992, it was a widely held agreement that farmers should be fully compensated for the income loss caused by the price cut. Over time, this item grew as institutional prices for other agricultural products had to be reduced in response to international pressure. However, the income loss incurred by farmers as a result of these further price cuts was only partly compensated by additional direct payments. Thus, the first group of farmers was treated better than subsequent groups.

Even if there was agreement that farmers had to be compensated, there was a widely held understanding that, a) it was compensation for the income loss caused by reduced institutional prices, and b) the compensation would diminish over time. This was based on the understanding that politically-set prices in the EU would continue to be higher than world market prices. The importance of the development of market prices can be illustrated using wheat as an example.

Figure 4 depicts the development of intervention and market prices for wheat and the attributed direct payments. It is obvious that market prices did not decline as much as intervention prices; however, the latter had been taken for quantification of the income loss. Moreover, market prices in recent years have been even higher than those that preceded the price cut. Most likely, in the coming years, prices will stay above the former price level (OECD-FAO, 2011). Furthermore, independent of the development of market prices, direct payments can no longer be justified by the need for adjustment aid. That part of EU agriculture (the Old Member States) which suffered from price reduction has had nearly 20 years of adjustment and thus, likely sufficient time to do so. It must be noted that a large part of EU agriculture - the New Member States - has never been hit by a price cut. Farmers in those countries generally enjoyed a higher income due to EU membership for ten New Member states in 2004, two more (Romania and Bulgaria) in 2007 and the last one (Croatia) in 2013.

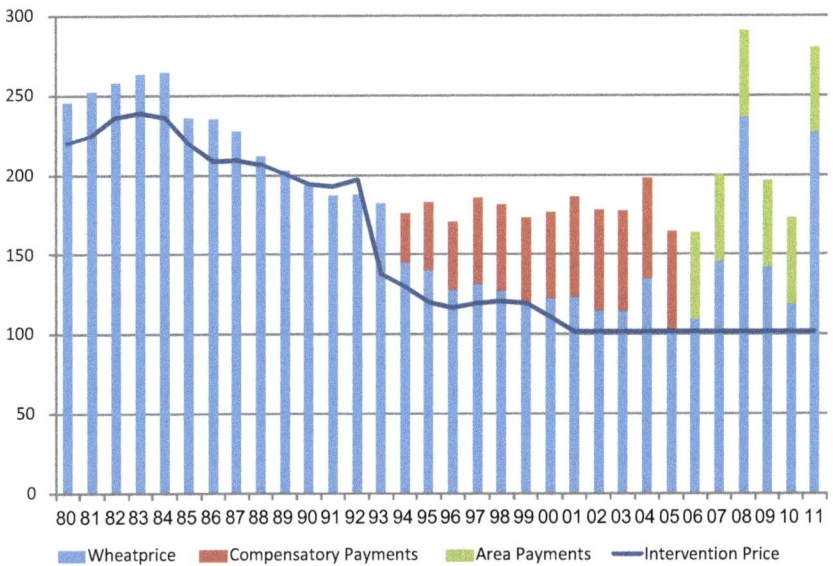

Fig. 4: Wheat prices and attributed direct payments in Euro per metric tons (year 1979/80 - 2010/11)

Source: Own calculation based on data of the European Commission

Despite changes in the economic environment and the length of the adjustment period, the Council of Agricultural Ministers and the European Parliament voted to keep approximately the same expenditure, but for a different purpose. Seventy percent of the payment per ha was designated as necessary for basic income payments and 30 percent for remunerating farmers for provision of environmental public goods. The evaluation of this policy change can be assessed based on the previously proposed criteria. Because the basic income support has different effects than the greening component (the 30% payment) the assessment must be done separately.

Basic income payments

Basic income payments could lead to an improvement of allocative efficiency only if the individual decision makers' actions were rated as more favourable from the society's point of view than would be the case without these payments, (e.g., if incentive compatibility would be improved). The assessment is based on the assumption that private agents (farmers) try to maximise profit and consumers try to maximise utility. Moreover, administrative costs, such as tax collection costs and disbursement costs must be taken into consideration.

Because payments are linked to area and independent from production amounts, a rationally behaving farmer will not change the production pattern. Payments do not alter the marginal revenue or the marginal costs for individual agricultural products. Of course, payments do lead to higher farm income and farmers who might otherwise close their farms without payments because of low income may continue to farm. These farms are generally less efficient than average. Closing less efficient farms would improve allocation of resources because a higher amount of land would be transferred to more efficient farms. Hence, one can conclude that allocative efficiency within the farm sector is negatively affected by these basic income payments. Additional costs will accrue for administration and collection of taxes, so taxpayers are the losers. Paying taxes lowers income and distorts the consumption pattern, giving rise to additional reductions in incentive compatibility.

There might, however, be one additional positive effect of basic income support[5] due to the cross-compliance rule. Farmers are only entitled to receive the full basic payments if they comply with environmental requirements which include two obligations. One concerns constraints for use of the land in agricultural production, (i.e., cross-compliance), and the other concerns the obligation to provide specific environmental goods. Concerning cross-compliance the European Court of Auditors (ECA, 2008) conducted a thorough investigation, and summarised its findings as follows:

- The objectives and scope of cross-compliance are not well defined, making it unclear what cross-compliance is designed to achieve.
- The legal framework poses significant difficulties, notably because it is too complex.

---

5   It might be argued that the society accepted this kind of support and, hence, policy makers have only responded to the will of the electorate. It is hardly believable that a well informed electorate would have voted for such a policy. There are many other distributional policies in the individual member countries but likely none with such an huge income transfer mainly to the better off population. Moreover, one has to keep in mind that social policy is still a matter of the member countries and not the EU.

- Cross-compliance and rural development are not well adapted to one another.
- Member States did not take seriously their responsibility to implement effective control and sanction systems. As a consequence, the control system provides insufficient assurance of farmer compliance.
- Data provided by the Member States on checks and infringements is not reliable and the Commission's monitoring performance was found "wanting" (ECA, 2008). The situation has likely deteriorated as the new member states which joined the EU after 2008 are less equipped to govern this instrument.

The three basic projected measures (EUROPEAN UNION, Press release, 26 June 2013) are:

- Maintaining permanent grassland and crop diversification (A farmer must cultivate at least two crops if his arable land exceeds 10 hectares, and at least three crops if his arable land exceeds 30 hectares. The main crop may cover, at most, 75% of the arable land, and the two main crops, at most, 95% of the arable area.).
- Maintaining an "ecological focus area" of at least 5% of the arable area of the holding for farms with areas larger than 15 hectares (excluding permanent grassland) – (i.e., field margins, hedges, trees, fallow land, landscape features, biotopes, buffer strips, afforested area). This figure will rise to 7% after a Commission Report in 2017 and a legislative proposal.

In principle, these measures could improve allocative efficiency, because they aim to reduce the divergence between private and social marginal costs and marginal benefits. However, first, these effects are supposed to be a by-product of basic income support and, hence, can hardly be seen as remuneration for producing environmental goods. Second, the payments are spread evenly across agricultural land, but the economic rationale for supporting production of environmental products differs significantly.

Generally, a measure which is not targeted is less efficient than those which are targeted. A measure can be targeted if, and only if, information on the actual situation is available and if the desired situation has been specified. Measures will be most efficiently implemented if they are focused as nearly as possible on the origin of divergence (CORDEN, 1997). The main measure for greening the CAP, the creation of ecological focus areas, places the same constraints on individual farms dependent only on size, where small farms are exempt from the constraints. However, there is very little doubt that the need for environmental improvements differs significantly across regions. Hence, the measure is neither targeted nor efficient. Moreover, it is not at all clear whether the environmental effects can be assessed, if not completely, at least within an acceptable margin of error.

In 2000, The European Court of Auditors concluded in its report (ECA, 2000), "However, neither the Community nor the Member States targeted the use of Community funds to the pre-established environmental priorities. Only five years later the Court (ECA, 2005) summarised its investigation as follows: "the verification of the agri-environmental measures poses particular problems and is far more resource-intensive than verification of the first pillar measure and indeed than other rural development measures. Such verification can rarely lead to even reasonable assurance at a reasonable cost." It can be assumed that this situation has not improved due to enlargement. The ECA found the main deficits in countries with domestic governance problems, and enlarging the EU has increased the number of countries with governance problems.

It is amazing that the Council and the European Parliament decided on these changes of the CAP in 2013 because the changes are not based on the evidence of past policies. The money is not well spent. Total welfare will most likely be lower than without (or with reduced) direct payments. Distributional effects are not in favour of those likely to have the lowest income and the greatest need. Such a policy change can hardly be welcomed by moral standards.

Summary

The aim of this article is to reveal the interdependence between morality and individual behaviour as postulated in mainstream neoclassical economics. It is emphasised that the term "morality" can be used in different ways, and any definition must include strong normative assumptions. Morality, as defined by deontologists, requires different behaviour than selfishly-oriented behaviour assumed by mainstream economics. However, if the definition of morality put forward by consequentialists is accepted, a happy marriage between morality and economics could be possible.

As the main element of morality, trust is very important for the functioning of markets, and markets are partly enforced by trust. However, markets as a means of coordinating individual decisions may lead to consequences which are not in line with a given society's moral code. The desired outcome will show up if, and only if, incentive compatibility prevails. It is argued that incentive compatibility is often not ensured under real world conditions. This holds true especially in agricultural and food markets. Products traded on these markets are primarily experience and credence goods as opposed to search goods. A strong government is needed to enforce incentive compatibility; however, the necessary institutional framework depends very heavily on incentives set in place by the institutional framework for policy decision making.

Market outcomes will be in line with the moral conduct code of the society under consideration only if incentive compatibility for political agents is secured. The article further offered structural suggestions for the needed institutional framework of EU decision making in agricultural and food policy. The last section elaborates on some recent policy decisions for reforming the Common Agricultural Policy from a morality perspective and how they must be rated.

List of References

Ahmed, M.N., Maas, S. and Schmitz, P.M. (2010): Analysing agricultural productivity growth in a framework of institutional quality. In: IAMO Forum 2010 Proceedings. Halle.

Baron, J. (1998): Trust, beliefs and morality. In: Ben-Neer, A. and L. Putterman (eds.), Economics, values, and organizations. Cambridge University Press.

Corden, W.M. (1997): Trade policy and economic welfare. University Press Oxford.

Debreu, G. (1952): A social equilibrium existence theorem. Cowles Commission for Research in economics, 38, 886-893.

European Commission (2013): Labelling of food stuffsfshttp://europa.eu/legislation_ summaries/consumers/product_labelling_and_packaging/co0019_en.htm.

ECA (2000): European Court of Auditors on "Greening the CAP" together with the Commission's replies. Special Report No 14.

ECA (2005): Concerning rural development: the verification of agri-environment expenditure, together with the Commission's replies. European Court of Auditors Special Report No. 3.

European Court of Auditors (2008): Is cross compliance an effective policy? Special Report No. 8.

European Union (2013): Press release, http://europa.eu/rapid/press-release_MEMO-13-937_en. htm.

Friedman, D. (2008): Morals and Markets. An Evolutionary account of the modern world. New York.

Gert, B. (2012): The Definition of Morality, The Stanford Encyclopedia of Philosophy (Fall 2012 Edition), E. N. Zalta (ed.), <http://plato.stanford.edu/archives/fall2012/entries /morality-definition/>.

Gilseman, M. B. (2011): Nutrition & health claims in the European Union: A regulatory overview. Food Science and Technology, 22 (10), 536 542.

Griffiths, B., Sirico, R. A. and Frankfield, N. B. (2001): Capitalism, Morality and Markets. The Institute of Economic Affairs, London.

Koester, U. (1977): The Redistributional Effects of the Common Agricultural Financial System. European Review of Agricultural Economics, 4 (4), 321-345.

Koester, U. (2014): Markets and Morality: The relevance for transforming the agricultural sector in transition countries. In: Kimhi, A. and Lerman, Z. (eds.), Agricultural Transition in Post-Soviet Europe and Central Asia after 20 Years, Studies on the Agricultural and Food Sector in Central and Eastern Europe, IAMO – Leibniz Institute of Agricultural Development in Central and Eastern Europe, Halle (forthcoming).

OECD and FAO (2011): OECD – FAO Agricultural Outlook 2011-2020. Rome and Paris.
Schmitz, P.M., Ahmed, M.N., Garvert, H. and Hesse, J.W. (2011): Agro-economic analysis of the use of Glyphosate in Germany. Agribusiness-Forschung, 28, Institut für Agribusiness, Giessen.
Schmitz, P.M. (1998): Das EU-Agribusiness im Globalisierungs- und Transformationsprozess. In: Herrmann, R., D. Kirschke and P. M. Schmitz, (eds.), Landwirtschaft in der Weltwirtschaft. Festschrift anlässlich des 60. Geburtstags von Ulrich Koester, Frankfurt a. M., 276-302.
Sedlacek, T. (2012): Die Ökonomie von Gut und Böse. München.
Sen, A. (2011): Die Idee der Gerechtigkeit, München.
SGS Deutschland (2014): SGS-Verbraucherstudie 2014: Vertrauen und Skepsis: Was leitet die Deutschen beim Lebensmitteleinkauf? http://www.lebensmittelklarheit.de/cps/rde/ xchg/ lebensmittelklarheit/hs.xsl/8705.htm
Smith, A. (1776): An inquiry into the nature and causes of the wealth of nations. London.
Smith, A. (1759): The Theory of Moral Sentiments. London.
Sysoyeva, M. (2011): Russia leaves 21% of arable and unused, according to Interfax http://www.bloomberg.com/news/2011-02-18/russia-leaves-21-of-arable-land-unused-according-to-interfax.html.
The Treaty of Rome (1957): http://ec.europa.eu/economy_finance/emu_history/documents/ treaties/rometreaty2.pdf.
Transparency International Deutschland e.V. (2013): Regulierung und Transparenz von Einflussnahme und Lobbyismus.
Transparency International (2011): Corruption in the Land Sector. Working Paper 4.
WISO (2014): Willkommen im Label-Dschungel. http://www.zdf.de/WISO/lebensmittel-label -siegel-supermarkt-bio-31473216.html.
Visser, O. and Spoor, M. (2010): Land Grabbing in Eastern Europe: Global Food security and land governance in post-soviet Eurasia. Paper prepared for presentation at the 118th seminar of the EAAE (European Association of Agricultural Economists), rural development: governance, policy design and delivery, Ljubljana, Slovenia.
Worth, M. (2013): Whistleblowing in Europe. Legal protection for whistle blowers in Europe. Transparency International.
Zühlsdorf, A., Nitzko, S. and Spiller, A. (2013): Ergebnisbericht Kennzeichnung und Aufmachung von Lebensmitteln aus Sicht der Verbraucher – Empirische Untersuchungsbefunde. www.lebensmittelklarheit.de/cps/rde/xbcr/lebensmittelklarheit/studiekennzeichnung-auf machung_ergebnisberich-2013.pdf.

Farm Animal Welfare: A Challenge for Markets and Policy

Monika HARTMANN, Johannes SIMONS and Kakuli DUTTA

Abstract

The public discussion about animal welfare in Western countries can be taken as an indicator for poor market performance regarding the implementation of a societal desired way of animal husbandry. From an economic point of view, governmental interventions to improve Farm Animal Welfare (FAW) can be justified in the case of asymmetric information, the existence of externalities or if animal welfare is regarded as a public or a merit good. All those market limitations seem to be given with respect to FAW.

An evaluation of policy intervention to enhance FAW standards needs to consider the interrelationship between national markets. Higher FAW standards have an impact on production costs and thereby on competitiveness. Divergent standards between countries may result in trade flows in favour of those countries with lower standards, thus harming producers in countries with stricter FAW regulations and limiting the potential positive effects for the welfare of farm animals. This holds, especially, as governmental action to prevent or limit imports of lower standard meat products and/or subsidy high animal welfare livestock products are restricted by international agreements.

The private sector has a greater scope of action and by that could enforce standards that are above the international level. The German example shows that the principles of such a private initiative are applicable. Nevertheless, a lot of contentious question have to be settled in order to be able to put these principles into action.

Acknowledgements

We would like to thank Klaus Frohberg for constructive comments. Any remaining errors are ours.

Introduction

While concerns about the appropriate treatment of farm animals used to be a discussion among a small group of activists it is now a highly debated issue in media, politics as well as among actors in the food chain in many Western countries. This growing concern is seen to have emerged as a response to advances in animal production technology over the last decades. Rising demand for meat products has led to an intensification and concentration of animal production, reducing production costs while at the same time increasing the challenges with respect to Farm Animal Welfare (FAW) (LASSEN et al., 2006; EDWARDS, 2004). Parallel to these developments, rapid urbanisation has enlarged the geographical as well as conceptual distance between consumers and producers and their respective animal production systems (EDWARDS, 2004; MÜLLER and SCHMITZ, 2002).

Many urban citizens have little first-hand experience with respect to the raising of farm animals. In fact, nowadays, urban citizens' knowledge regarding animals is dominated by their relationships with pets or their visits to the zoo (EDWARDS, 2004). In addition, the rising relevance of FAW can be attributed to significant shifts in values and preferences of an affluent society especially in Western countries (e.g. OLYNK, 2012; NAPOLITANO et al., 2010).

Assuming well functioning markets a higher valuation for a product (characteristic) should be reflected in a higher Willingness to Pay (WTP) of (some) consumers for that product (characteristic), thus stimulating production for the desired good. Regarding animal welfare, however, several challenges exist for markets to deliver the optimal outcome: First, animal welfare is a complex concept and has proven difficult to define and to assess even at the level of experts. Second, also if the analysis is confined to what we (as human beings) perceive as FAW, it opens up challenges for the market. FAW is a credence attribute which is not easily verifiable before or after purchase. To substantiate the truthfulness of information on animal welfare is difficult if not impossible due to considerable or even prohibitive high information costs, thereby leaving considerable room for fraud. Market forces (e.g. third party certified labels) might be able to reduce the existing information asymmetry thereby preventing fraudulent claims. Nevertheless, it seems likely that markets still will not deliver an optimal outcome with respect to FAW due to its non-private good characteristics (NORWOOD and LUSK, 2011). This third challenge raises the question whether governmental regulations are required and, if so, how as well as to what extent potential trade-offs in a global market need to be taken into account.

This paper conceptually analyses farm animal welfare from an economic point of view. We start out by discussing the ambiguities associated with the definition and the assessment of FAW. This is followed by investigating the perception of animal husbandry and production from the point of view of producers and consumers. Given the market challenges in securing FAW, brings us to discussing why and how governments can intervene pointing also to trade-offs of government intervention. Given those limitations we suggest a new role for markets linking these ideas to recent developments in Germany. The paper concludes with a brief summary of the main findings.

Ambiguities with respect to farm animal welfare

Ambiguity with respect to definition and assessment of FAW
Despite the relevance of FAW for the public there exists no commonly agreed upon definition that could assist in the development of an integrated assessment methodology for evaluating FAW (SWANSON et al., 2011; CRONEY and MILLMAN, 2007; FRASER et al. 1997). In the following we discuss three contrasting schools of thought that each tries to substantiate the abstract idea of FAW. However, given their different orientation they also contribute to the disagreement associated with respect to FAW (FRASER et al., 1997; HEWSON 2003; CRONEY and MILLMAN, 2007).

(1) According to the concept of 'biological functioning' animal welfare is linked to the absence of large physiological stress response, the capability of an animal to adapt to its environment and its ability to satisfy its biological needs. Thus a broad spectrum of the behavioural, physiological, health and fitness responses of animals are considered appropriate indicators to evaluate FAW. (2) The feelings-based perspective defines animal welfare in terms of absence of adverse emotions such as pain, fear and frustration, and the enhancement of positive emotions such as comfort and pleasure (DUNCAN, 2005; DUNCAN and DAWKINS, 1983;). (3) The 'natural behaviour' concept equates FAW with animals' freedom to express natural behaviour (SPINKA, 2006).

Those three concepts and the derived indicators are controversially discussed in the literature (e.g. DUNCAN, 2005). The school of biological functioning focuses on an animal's health and its capability to cope with its environment, e.g. measured by e.g. endorphins, cortisol, and heart rate (HEWSON, 2003). As it is assumed that an animal is only productive and efficiently reproducing if it is healthy, this concept is closely linked to economic parameters such as e.g. productivity indicated by number of piglets per sow or milk yield per cow. Those involved in the sector acknowledge the biological functioning concept as especially relevant in defining FAW (see discussion below).

The overall critique on animal husbandry demonstrates that societal groups contend this approach, emphasizing on the importance of an animal's mental state. According to these groups, physical parameters (such as heart rate) are difficult to interpret, because they change based on positive and negative experiences, such as the presence of a mate or of a predator. This appears to validate the feelings–based concept which, however, faces other challenges. Feelings are characterized by subjectivity, are vaguely defined, thus impossible to measure directly, and difficult to determine indirectly (DUNCAN, 2005). Nevertheless, scientists have developed laboratory techniques derived from cognitive psychology such as preference tests, to assess FAW from the feelings-based perspective (for an example see TUCKER et al., 2003; for an overview of relevant literature see DUNCAN, 2005). These tests are based on the assumption that the animal will choose options according to how it feels, that is, in the best interest of its welfare. However, it has been debated that preference tests may e.g. be affected by previous experiences of the animals and thus give only relative information, focus on current feelings of the animals without considering the long-term consequences on their welfare and neglect the interaction of different determinants of FAW to which animals are exposed to in the real world (DUNCAN, 2005; CRONEY and MILLMAN, 2007). For those reasons, feelings-based indicators need also be treated with caution. The third school of thought emphasizes the freedom of the animal to express natural behaviour. According to SPINKA (2006) such behaviour may occur in efforts made by animals for survival or for coping with extreme adverse situations. Though without any doubt this behaviour is a natural response, it has its deficiencies in measuring FAW because domesticated animals are not exposed to those situations.

As is evident from this brief discussion, FAW reflects a wide range of animal-related considerations and leads to different evaluations depending on which of these three concepts is adhered to (LASSEN et al., 2006). In view of that, the same husbandry practices may be judged differently: Sows in cages lack free moving space but are well fed and efficient regarding reproduction. According to the biological functioning approach this indicates a high level of animal welfare. But welfare is compromised if we evaluate such practices based on feelings or natural behaviour because these sows are prevented from expressing natural behaviour (FRASER et al., 1997).[1]

The exposition reveals that animal science has developed methods to better understand the impact of rearing conditions on the physical, physiological and the well being aspects of FAW. In light of the above debate, however, it has to be acknowledged that the concept of FAW and related to it the methodologies for measuring FAW remain contentious.

Ambiguity with respect to the perception of FAW

Controversies regarding definition and measurement are, however, not the only challenge with respect to FAW. Even more demanding is the ethical decision where to set the limits with respect to still acceptable FAW and where mistreatment begins (ESTEVEZ, 2003). Along the same lines and closely related, answers

---
[1] See also the discussion on egg production in cage-free versus cage systems (NORWOOD and LUSK, 2011).

have to be found for other ethical questions, such as 'Are today's animal rearing conditions legitimate? And to what extent are compromises necessary in a less-than-perfect world? And are we as consumers or citizens willing to pay the price for higher animal welfare standards (LASSEN et al., 2006; ESTEVEZ, 2003)?'

These questions point to the ethical and economic dimension of FAW. They reveal the relevance of human perception for securing FAW. Based on this perception we take concerned actions because as individuals we identify ourselves as ethical human beings. Since we speak on behalf of the animal, animal welfare is in reality a "subset of human welfare, the animals' preferences and wellbeing having relevance only to the extent that they are important to us." (MCINERNEY, 2004). Thus, "if there is any conflict between our preference and the animal's preference, it is ours which inevitably prevails." (MCINERNEY, 2004). In fact our perception of what is acceptable may vary considerably depending on the animal. For instance, public outrage is strong when knowledge of puppy mills is made available, though the conditions for farm animals might be similar or even worse (NORWOOD and LUSK, 2011). Even at the level of farms the status between the categories of "farm animal" and "pet" can be rather blurred. Farmers often develop feelings of friendship with the animals kept in their vicinity for longer periods but at the end of the production process, such feelings vanish or are suppressed as the animals have to be "recommodified" (WILKIE, 2005).

The ethics of animal welfare depends on humans and the relation of humans and animals. And: If FAW conflicts with other objectives, as has been the case with respect to the Asian Flu or BSE (Bovine spongiforme Enzephalopathie) it is regarded as justified to kill thousands of animals.

FAW and Market Challenges

In market oriented economies consumers make decisions to maximize their preferences thereby influencing what is produced and offered on the market. Products will not be offered for which the costs of production are not covered by the WTP of (at least some) consumers. Thus, it is important to investigate the role and tradeoffs of FAW at the producer and consumer stage.

Producer perception of FAW and existing trade-offs

Producers primarily associate FAW with basic biological needs, the first FAW concept discussed above (BOCK and HUIK, 2007; TE VELDE et al., 2002). As producers are physically closest to the animals they are breeding, housing and feeding, they develop a general tendency to consider themselves as knowledgeable and rational actors and undermine the concerns of the consumers and other stakeholders (animal activists) as biased and uninformed (VAN HONACKER et al. 2007). Commenting on the heterogeneity of producers regarding their perception of FAW, BOCK and VAN HUIK (2007) in their cross country EU study pointed out two contrasting groups of farmers. The first group conceived FAW to be related to animal health and the basic physiological needs and perceived good FAW to be important for economic reasons. Farmers of the second group[2], associated FAW primarily in terms of the expression of natural behaviour. Though animal health was as well seen to be important for economic reasons, securing animal welfare above and beyond the biological functioning was motivated by ethical concerns (BOCK and VAN HUIK, 2007).

The different positions of farmers and their respective production systems can be illustrated by the welfare-productivity frontier in Figure 1 which conceptually shows the trade-offs between Total Livestock productivity (TLP) and Perceived FAW (MCINERNEY, 2004). Within each livestock sector (e.g. dairy sector, poultry sector) production systems are distributed at different points on and inside the frontier depending on their technical efficiency and intensity. The concave nature of the frontier reflects that at high levels of FAW productivity gains can be realized with only small compromises regarding the welfare of farm animals. However, at higher productivity levels further gains in productivity imply high costs in terms of FAW (MCINERNEY, 2004). Accordingly, the marginal rate of substitution (MRS) in production between FAW and TLF increases with productivity (see Figure 1). If factor prices do not change with an increase in TLP we can assume a negative relation between costs of livestock production and TLP. Thus, farmers who primarily focus on productivity produce close to point F, implying low costs and a modest degree of FAW. Others placing relatively more value on FAW might select e.g. point C. Assuming highly competitive markets the realisation of ethical preferences (e.g. point C compared to point F in Figure 1) is in the long run only possible to the extent that the higher costs due to the reduction in TLP (distance x) is compensated by consumers due to a higher WTP for the increase in Perceived FAW (distance y). This would imply that farmers are able to reap the economic benefits of a higher price niche market and thereby remain competitive.

The conceptual model in Figure 1 reveals, that, other things being equal for a competitive and efficient industry, there exists a trade-off between improving farm animal welfare and productivity from point A onward (MCINERNEY, 2004). However, there are three exceptions: a) over-intensification which lead to

---

2   Groups were formed according to their participation in specific animal welfare or organic production schemes. Thus, all farmers in the second group participated in a respective scheme, while this did not hold for members of the first group.

a breakdown of the system and, thus, to a reduction in both TLP and FAW (going beyond point F in Figure 1); b) the introduction of new technologies, genetic progress, better inputs and others, which move the frontier outwards, can advance both, FAW and TLP; c) enhancing the performance of the less efficient livestock producer in the sector (e.g. at point G) can result in improvements of animal welfare (from G to B), of productivity (from G to C) or any combination of both (any point between B and C on the frontier). Accordingly, investment in R&D as well as identification of laggards and supporting them in approaching the welfare-productivity frontier are win-win possibilities for FAW and TLP (HARVEY and HUBBARD, 2013).

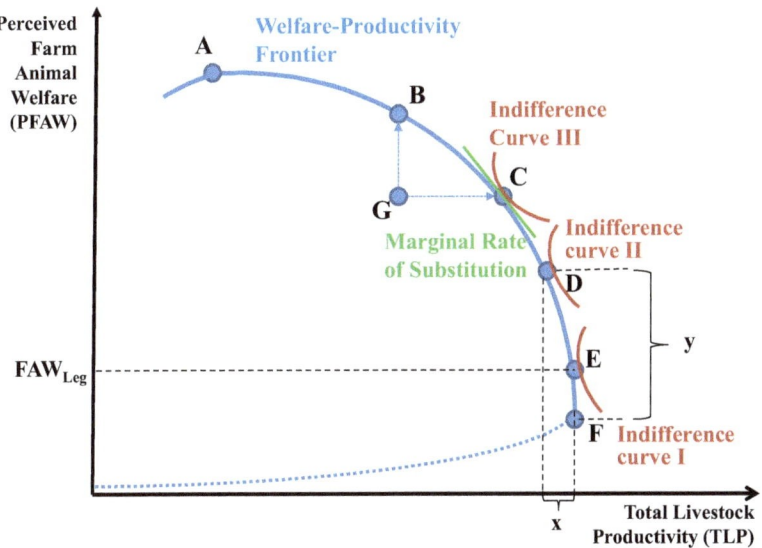

Fig. 1:   *Welfare and Productivity Frontier: Conceptual Framework*
Source: Adapted from MCINERNEY, 2004

FAW legislation can be considered as a force stimulating farmers to move upwards along the frontier. But, the question is who will bear the burden of improved farm animal welfare (e.g. a move from F to E in Figure 1)? Higher standards for FAW cause adjustment costs primarily at the farm level. Farm production costs are just one part of the total costs that add up to the price of the product at the retail level. Thus, it will depend on the supply and demand elasticity and the market power of the respective economic agents along the chain (farmers, processors, wholesalers, retailers and consumers) who will bear most of those costs (NORWOOD and LUSK, 2011). In this respect it can be assumed that the farmers on the one end of the chain and the consumers on the other end are the weakest links.[3] Higher retail prices of meat might reduce meat consumption or encourage consumers to search for imported meat products with a lower price but also a lower FAW standard. Higher costs at the level of farmers if not reflected in higher product prices might drive some producers out of business, thereby reducing domestic production. Thus, FAW standards might be considered as 'gold plating', even if well-intended, if they are critical for the survival of domestic farmers and potentially induce cheating in the value chain (SWANSON, 2008).

Thus, the question is how to determine the social utility maximising point on the welfare-productivity frontier. It is very likely that the economically optimal level of FAW and TLP varies between societies. The optimal point with respect to the level of animal welfare and productivity for a society is given where the MRS in production is equal to the MRS in consumption and thus, where consumers' welfare indifference curve is tangential to producers' welfare-productivity frontier. Richer and better educated societies are more able and likely more willing to pay for improved animal welfare. In the terms of Figure 1 this implies that the marginal rate of substitution in consumption between FAW and TFP is small for affluent societies: the loss in animal welfare they are willing to accept in return for some gain in productivity is small (Point C in Figure 1). In societies where animal welfare is of no or marginal importance, livestock production systems tend to be driven towards maximum productivity (point F in Figure 1). Thus, a closer look at consumers' perception of and WTP for FAW is of relevance for the discussion in this paper and will be considered next.

With few exceptions there exists a trade-off between FAW and TLP. This implies that rising standards of FAW come at a cost. The optimal point with respect to the level of FAW and TLP for a society is determined by the costs of securing FAW in terms of induced productivity losses and society's values regarding FAW relative to TLP.

---

3 The relative change in prices at the retail level for a given rise in production costs due to higher FAW standards depends in addition on the degree of transformation of the farm product to the consumer level (MCINERNEY, 2004).

Consumers' and citizens' perception of FAW and their WTP

Consumers/citizens see current rearing conditions more critical than farmers (e.g. VANHONACKER et al. 2008; MÜLLER and SCHMITZ, 2002).[4] The Eurobarometer on 'Attitudes of EU citizens towards Animal Welfare' (EU COMMISSION, 2007) indicates that FAW is considered to be of great importance (Average score 7.8 on a scale from 1 'not at all important' to 10 'very important'). In addition, the vast majority of participants in this EU wide survey believes that welfare protection of farm animals in their respective countries needs to be improved (35% of the respondents 'certainly needed', 42% 'probably needed') (EU COMMISSION, 2007). Those opinions are formed despite the fact that the majority of Europeans (57%) declare to know only 'a little' and 28% even say that they understand 'nothing at all' about the conditions of animal husbandry in their country. Only 12 % of the respondents claim to be informed 'a lot' (EU COMMISSION, 2007).[5] Furthermore, multiple studies have shown that consumers have a (high) WTP for improved FAW (see studies cited in NORWOOD and LUSK, 2011 and the meta-analysis in LAGERKVIST and HESS, 2011).

One would assume that people's high interest in and WTP for FAW is reflected in the market place. In reality, however, the market for animal-friendly products still resembles not more than a niche (NORWOOD and LUSK, 2011; INGENBLEEK et al., 2013). This raises the question, 'Are the results obtained from research studies suitable in revealing consumers' market preferences?' Most studies analysing consumers' WTP for FAW request consumers to make trade-offs between e.g. price, convenience and ethical attributes such as FAW. Nevertheless, many analyses are of a hypothetical nature, and thus the results are prone to suffer from the abstractness of the situation for consumers (e.g. LAGERKVIST and HESS, 2011). In addition to this hypothetical bias, the results are likely influenced by a socially desirability bias. Given this issue investigated responses probably overstate the relevance of FAW for consumers 'real' purchase decisions (ALPHONCE et al., 2013; HARTMANN, 2011). Thus, the obtained findings are closer, though probably not identical[6], to what consumers' feel how they 'should behave' as citizens and not how they indeed 'behave' in the market. In the following we refer to the former as reflective preferences and to the latter as market

---

4 They also have a somewhat different perception about what constitutes good FAW (e.g. animals' ability to express natural behaviour is of greater relevance to consumers/citizens than to farmers) (Vanhonacker et al. 2008; MÜLLER and SCHMITZ, 2002).
5 Information obtained by media is the main sources on which consumers' and citizens' base their evaluation of FAW (MÜLLER and SCHMITZ, 2002).
6 Reflective preferences should be independent of ability to pay and thus income. However, LAGERKVIST and HESS (2011) show in their meta-analysis that respondents' income has a significant positive impact on WTP.

preferences. The consumer-citizen duality or consumer-citizen gap (see also the discussion below) can be linked to the contrasting roles individuals have as consumers and as citizens when it comes to meat production and consumption (e.g. GRUNERT, 2006; VANHONACKER et al., 2007; ALPHONCE et al., 2013). However, this just leads to a corresponding question: 'Why do preferences revealed in the market place by an individual's purchase behaviour with respect to meat differ from an individual's attitude towards the process of producing those types of meat?'

In the literature several barriers are identified that prevent individuals from translating their reported ethical attitudes into behaviour. Those include a) no availability to or high transaction costs in obtaining respective products, b) insufficient or unreliable information with respect to such goods, c) the price of animal welfare friendly products, and d) perceived lack of effectiveness in as well as e) responsibility for securing FAW (see e.g. TOMA et al., 2011).

As regards a) lack of availability of animal welfare friendly products or high transaction costs to obtain such products there is some ambiguity involved; at least in the medium to long run it is not a cause but rather a consequence of some other reasons to be discussed below. This holds if one can agree on the general assumption of a market oriented economy that consumers influence with their buying decisions what is produced and offered at the market.

Moreover, b) information was identified as another potential barrier for purchasing animal welfare friendly products. According to the Eurobarometer on animal welfare the majority of EU consumers/citizens would like to have more information on FAW (EU COMMISSION, 2007). In addition, the survey reveals that in comparison to respondents stating to know little about animal husbandry those who perceive themselves to be quite knowledgeable with respect to FAW attach a higher importance to the well-being of farm animals and like to see farmers be compensated for improvements in this respect (EU COMMISSION, 2007). Based on a structural equation model TOMA et al. (2011) found that access to information about conditions under which animals are raised has a positive significant direct impact on attitudes as regards to FAW and an indirect one on animal welfare-friendly consumption behaviour. Along the same lines LAGERKVIST and HESS (2011) derive from their meta-analysis the conclusion that information on FAW supports consumers in forming preferences for improvements in FAW. However, providing information is, firstly, not without problems given the complexity and ambiguity of the concept of FAW discussed above and, secondly, likely not sufficient because of the credence nature of FAW. In this context the underlying information asymmetry becomes an issue. Purchasing animal welfare friendly labelled meat a consumer must trust that this originates from an animal which was raised in a responsible manner. Thus, the strength of e.g. an animal welfare label depends on its credibility which, in turn, is influenced by several factors. To those belong i) competence and independence of the institution granting the label, ii) transparency of the criteria used (which, as already discussed, are not easily defined), iii) control of their compliance and iv) the extent to which stakeholders are knowledgeable about those issues (HARTMANN, 2011). As IN-GENBLEEK et al. (2013) show third party certified FAW labelling schemes exist throughout Europe. However, the respective market is presently highly fragmented and in most countries the market share of those products is low (see the discussion above).

Another barrier to translating ethical attitudes into behaviour is c) the price of animal welfare friendly products. Even if consumers are willing to pay more for those products, the mark up in the supply chain may go beyond this WTP leading to the well-known market gap (to be discussed below). Indeed NORWOOD and LUSK (2011) show for the United States that the premium for cage free eggs compared to cage eggs at the consumer level is considerable (on average more than 120% over the period 2004 to 2008) and thereby exceeds the WTP estimated in most studies. Interestingly, the mark up for animal welfare products identified in the study of NORWOOD and LUSK (2011) cannot exclusively be explained by higher production and distribution costs, but is partly linked to higher margins gained by the distribution sector.

The fourth and fifth barriers refer to d) the perceived lack of effectiveness and e) responsibility with respect to FAW and thus are closely linked to the question whether FAW is a private, merit or public good and thus whether the government is to become active in securing farm animal welfare.

The high value consumers express for FAW is not reflected in the market place. Animal welfare friendly products are still a niche market. A number of barriers can be identified that help explain why consumers do not 'walk their talk'. Some of those barriers point to market failure and thus the need for governmental intervention.

FAW and the Policy Challenge

Justification for Government Intervention

Analyses of market oriented economies primarily rely on the concept of private goods which is directly linked to the assumption of the existence of well defined consumer preferences driving the market. However, as discussed above the existence of information asymmetry with respect to FAW might lead to market failure deeming governmental intervention necessary. In this case public policies such as education programs and providing or requesting the provision of information at the POS (Point of Sale), e.g. by labels, can help to overcome the shortcomings of market forces. In this respect, it needs to be considered that consumers are very heterogeneous regarding their capability and willingness to process information (VERBEKE, 2005). For some groups of consumers the opportunity costs of processing (extended) animal welfare related information is too high compared with the derived marginal benefits. Some might be reluctant to consider respective information due to the complexity of the issue, while others might not want to be confronted and have to cope with the topic 'FAW' as it probably will activate feelings of unease and could negatively affect the pleasure of eating meat. Accordingly, VERBEKE (2005) suggests providing targeted information to the needs of different segment of consumers. The need to overcome the existent information asymmetry has also been realised by the EU COMMISSION. To support interested consumers in identifying and choosing welfare-friendly products the EU Commission has been examining the introduction of 'reserved term' based on farming methods or of standardised welfare indicators on a mandatory or voluntary basis (EU COMMISSION, 2009).

Market failure can, however, also occur due to the existence of factual constraints – non-rivalry in consumption and impossibility of exclusion – that prevents individuals from acquiring goods (public goods) or preventing bads (pubic bads) to the extent they actually desire. Assuming that those goods (bads) are of significance to almost everybody provides a justification for government intervention. The situation is different with respect to the concept of (de)merit goods first introduced by MUSGRAVE (1956). Merit wants are "those which individuals may not be willing to express by their overt actions, but which informed leaders of the social group consider important enough to be deliberately encouraged and promoted." (MUSGRAVE (1959), cited in VER EECKE, 1998, 136). For those goods there are no factual constraints. Their existence seems to rely on value judgment of authorities (VER EECKE, 1998). As they are in conflict with (revealed) individual preferences[7] they challenge the fundamentals of the traditional economic model and have been suspected of "representing the first step on the slippery slope of paternalism" (D'AMICO, 2009). This has made it difficult to defend them (KIRCHGÄSSNER, 2012; BESLEY, 1988). Merit goods, however, find their economic justification if one acknowledges multiple utility orders (e.g. KIRCHGÄSSNER, 2012; CAMERER et al., 2003; MANN and GAIRING, 2012). Findings of behavioural economics are indicative of the fact that individuals' market preferences are often short term oriented and can deviate from their

---

[7] This is a difference compared to public goods (bads) where consumers' wishes are acknowledged, though not directly, by market forces.

higher order or reflective preferences discussed above (e.g. SUNSTAIN and THALER, 2003). In case of divergence (consumer-citizen gap) between markets and reflective preferences, political processes offer society the possibility of self-binding. Thus, in a democratic legitimized system it can be argued that the provision of merit goods or the prevention of demerit ones is backed by citizens' political support. For consumers it can well be rational to vote in favour of their 'reflective' preferences which might or might not be revealed by their market behaviour (e.g. KIRCHGÄSSNER, 2012; MANN, 2003).

FAW, as many characteristics or goods, exhibits elements of all three concepts (MCINERNEY, 2004; VER EECKE, 1998). For illustrative purpose, suppose that the only reason consumers assign a high value to FAW is their belief that it leads to better-tasting meat. According to this argument FAW can be considered as a private good for it can be provided by markets for animal welfare friendly meat and purchased by consumers (at higher prices) to the desirable extent. Now assume that improved FAW is valued purely out of concern for the farm animals and that perfect information is given. If consumers derive e.g. a higher pleasure of eating meat from animals knowing they were appropriately raised their WTP for that good is higher compared to low standard products. Also in this case, consumers could decide for animal welfare friendly products provided by markets.[8] However, consumers likely do not exclusively care about the well-being of the livestock whose meat they later eat. And vegans, though not at all participating in the market for livestock products, may considerably suffer if farm animals are treated in a cruel way (MANN and GAIRING, 2012). Thus, the costs related to bad FAW or the benefits linked to superior FAW are externalities of livestock production and consumption (MCINERNEY, 2004; NORWOOD and LUSK, 2011). As shown by the Coase theorem, externalities can be mitigated by private party negotiations if trade in these externalities is possible, property rights well defined and transaction costs low (MEDEMA, 2014). Based on citizens' WTP for various levels of FAW and producers costs of securing them a market for FAW would need to develop separately from the market for meat.

However, FAW is non-rival in consumption (once it is secured for one person (some of) the benefits can be enjoyed by all those caring about farm animals and it is non-excludable as no-one can be prevented from enjoying better animal care (HARVEY and HUBBARD, 2013; MCINERNEY, 2004). Due to the non-payer non-excludability and the resulting free rider problem (people can enjoy a higher level of FAW without paying for it) and independent, whether it is traded in a separate FAW market or as presently linked to the meat market, FAW is bound to

---

8 Though, there only exists a market for animal friendly meat if the additional WTP of consumers is at least as high to cover the extra costs due to improvement of FAW.

market failure. Thus, consumers may feel that with their purchase decisions they will gain little impact on the overall well-being of farm animals, leading to the perceived lack of effectiveness. This, however, makes consumers also more prone to delegate the responsibility for securing FAW to others, e.g. retailers or the government (TE VELDE et al., 2002).

The 'free rider problem' could be used as an argument for governments to secure FAW on behalf of society. According to the public good argument intervention would improve societal welfare if the 'free rider gap' with respect to a specific level of FAW is greater than the 'market gap' (HARVEY and HUBBARD, 2013). The former is defined as the difference between citizens' WTP assuming a world without free riders (thus, everybody can be sure that all others who value FAW also pay for it) and citizens' revealed WTP in the market (vertical distance between curve WTP$_{FAW,No\ free\ riders}$ and curve WTP$_{FAW,\ revealed}$ in Figure 2). The 'market gap' is equal to the costs necessary to secure a specific level of FAW in the market and citizens' actual WTP as revealed in the market (vertical distance between curve Cost$_{FAW}$ and curve WTP$_{FAW,\ revealed}$). Figure 2 reveals that up to the level FAW$_1$ the 'free rider gap' exceeds the 'market gap', implying that societal welfare could be improved by government intervention.

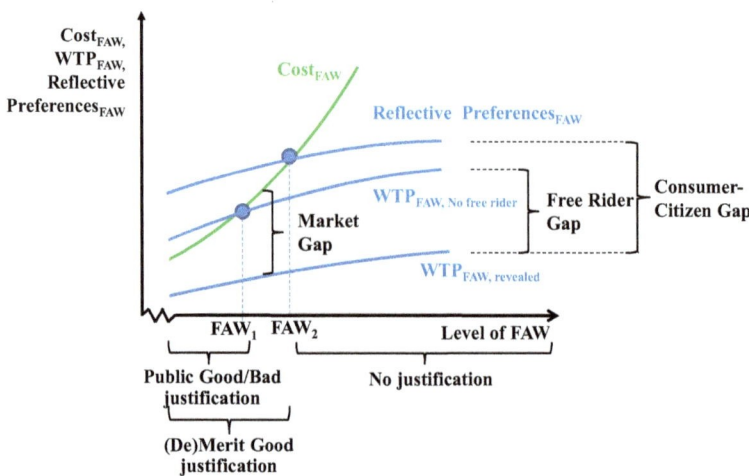

Fig. 2:   *The public and merit good justification for governmental intervention*
Source: Own presentation

The public good case for intervention depends on citizens' ability to pay and on their preferences for FAW as revealed in their WTP on the one hand and the level of FAW on the other. Especially for animal husbandries that are characterized by varying degrees of neglect, cruelty and uncaring management it is likely that the 'free rider gap' exceeds the cost of preventing such treatment at the farm level up to the slaughter house (MCINERNEY, 2004). This would make low levels of FAW a public bad. At higher levels of FAW the 'free rider gap' only slowly rises with animal welfare improvement, while the respective costs strongly increase, thus eliminating the public good case for governmental intervention (see Figure 2).

The merit good argument comes into play when a gap between consumers' market preferences and their reflective preferences with respect to FAW exists. It also rests on the value judgment that the latter reflect citizens 'true' preferences. The 'consumer-citizen gap' or "consumer-citizen duality" (VANHONACKER et al., 2007) equals the difference between the perceived value of FAW (which is independent of consumers' ability to pay for FAW) and citizens' WTP as revealed by the market (vertical distance between curve Reflective Preferences$_{FAW}$ and curve WTP$_{FAW, revealed}$). Thus, even if cruel treatment would fail the 'public bad' test for justifying governmental intervention as discussed above it might pass the 'demerit' test. Along the same lines a higher level of FAW (levels between FAW$_1$ and FAW$_2$), might not be justified based on the public good discussion but may so according to the merit good argument. An interesting case in this respect is the 2008 California ballot proposition on animal welfare. 63.5% of Californians voted e.g. for the proposal to ban cage eggs though, at the time, the market share of cage-free eggs was well below 5% (ALPHONCE et al., 2013; NORWOOD and LUSK, 2011). However, in case where there is no ballot by the population (and assuming no manipulation prior to the voting) the problem of measuring reflective preferences remains a challenge. As a consequence, the decision on the level of FAW is beyond the scope of economists and falls to a large extent into the political arena. Thus, it is prone to policy failure.

FAW has characteristics of a private, a public and a merit good. To what extent government intervention is needed depends on citizens' ability and their willingness to pay for FAW, their reflective preferences with respect to FAW as well as the costs to secure FAW. Though it is a challenge to measure all these characteristics special difficulties arise with respect to citizens' reflective preferences. To prevent policy failure 'soft' interventions are desirable.

Limitations of governmental intervention

Society signals its political preferences for the appropriate mixture and level of farm animal welfare which are (to some degree) reflected in the country's laws and regulations given by FAW_{Leg} (see Figure 1). Such standards set by legislation might well deviate from consumers' revealed preferences for FAW as discussed above. However, it might as well lag behind societies reflective preferences (possibly point D in Figure 1) due to lobbying of interest groups or external constrains (see below). Regarding the economic consequences of FAW legislation at the example of the EU one has to differentiate between (1) higher standards in one or several of the EU member states[9] and (2) higher standards for all of them. [10]

Increasing standards for animal welfare in just a few of the member states leadv to additional costs for respective producers in these countries. As it is not possible to build up trade barriers between EU member states or to subsidise production of meat in those countries, higher standards result in a competitive disadvantage for domestic producers. As a consequence production in the respective member states will decrease in favour of the other ones. These detrimental effects are the stronger the higher the costs linked to the new legislation, the lower the share of consumers' willing to pay a higher price for the domestically produced animal welfare friendlier products and the lower the difference in transaction costs between domestic and foreign products. It is obvious that in this case also the aim to improve FAW will be accomplished only to a lower degree if at all.

Implementing higher FAW standards in all EU member states leads to a different situation. FAW standards existing already today in the EU are in an international comparison high (NORWOOD and LUSK, 2011). But, at present, European meat markets can be highly protected by tariffs in line with the WTO Agreement on Agriculture (WTO 1995a) and the Goods Schedules annexed to the Marrakesh Protocol to this agreement (WTO, 2013b). Accordingly, a search for imports so far did not and likely will not occur following a (moderate) increase in FAW standards. However, in case of tariff reductions or even their abolition, the protection of the domestic markets would decrease or vanish. Increasing FAW standards in the EU would then ease market entry from countries with low or even

---

9 The same problems discussed below arise in e.g. the US as farm animal welfare legislation is decided on at the level of the state. Thus, different standards between US states exist with their resulting impact on inter-state trade (see NORWOOD and LUSK, 2011).
10 The impact would be again different, if we assume a closed economy. Though this would imply that consumers lack the option to substitute higher priced animal welfare friendly products by lower price lower standard once, farmers are nevertheless affected. The higher the increase in consumer prices due to higher FAW standards and the higher the price elasticity of consumption the stronger is the fall in demand and given the closed economy assumption, accordingly, in supply of livestock products.

without any FAW standards while reducing EU animal production thereby potentially leading to a decrease in overall FAW.

Implementing trade barriers based on process standards (higher FAW standards) is not allowed according to the SPS agreement of the WTO if the standards have no impact on the physical product characteristics. And even if those standards have such an impact, trade barriers are only justified if a negative impact on human health is based on scientific evidence (WTO, 1995b).

Regardless of the protection of domestic markets by tariffs or non tariff barriers, the competitive situation of European exporters of animal products into third country markets would worsen, if the gap between the additional costs due to higher FAW standards were not covered by export subsidies. It is the WTO that sets the frame for export subsidies. In December 2013 WTO members agreed on abolishing export subsidies so that in the medium term it will not be allowed to balance higher costs due to animal welfare standards by granting export subsidies (WTO, 2013a).

Speaking about the WTO one has to keep in mind that its rules refer to state activities. This implies that though the EU cannot require importers into its member states to fulfil standards of FAW exceeding agreed international ones private persons and private companies can do that. Thus, e.g. retailers are free to only buy products that fulfil their own FAW standards which are potentially higher than the ones set by governments. Neither EU firms nor EU consumers can be accused if they refuse buying goods that are not in line with those higher standards. Taking this into account the question is, 'Is it possible to increase FAW standards via markets'.

The EU and its member states are linked to world markets by trade flows. Increasing FAW standards above the international level reduces the competitiveness of EU farmers as international agreements limit the scope of the EU to compensate farmers for higher production costs. Standards leading to an increase in production in 'animal welfare hells' at the expense of EU production reduce FAW.

A new role of market participants

Though food processors and retailers are free to only source products meeting higher standards with respect to FAW, they face conflicting interests. On the one hand, such a strategy supports a reputation of responsibility and thus likely increases consumers' confidence, trust and loyalty in the retailer (HARTMANN, 2011). On the other hand, demanding more efforts from suppliers with respect to FAW is, as discussed above, linked to an increase in costs at the producer level. Accordingly, it needs to be adequately compensated. Otherwise those requesting higher FAW standards will lose their suppliers to their low standard competitors. Paying, however, higher farm prices will result in a mark-up on prices at the consumer level and likely in a loss in market shares in the consumer market. The latter holds as the chance to enforce standards based on consumers' choice of goods is difficult and at present primarily applicable for niche markets as was explained above (see section 'Consumers' and citizens' perception of FAW and their WTP).

Thus, market based solutions seem to need some coordination among competitors in the chain to be viable. One interesting attempt in this respect is the "Initiative Animal Welfare" (Initiative Tierwohl) launched by all relevant actors in the German meat chain (QS, 2013). The scheme is being developed out of the QS-System, a cross-stage (production, processing and distribution) and cross-company quality assurance scheme for meat and meat products as well as other fresh products which has been implemented in Germany after the BSE crisis (QS, 2014).

The aim of the 'Initiative Animal Welfare' is to develop in a joint effort of all actors of the meat chain measures to improve the welfare of the animals in the pork and poultry sector. The strategy for the pig sector can be characterized as follows (QS, 2013):[11]

- Retailers pay into a fund depending on the amount of QS labelled pork meat and meat products that they sell in their outlets.
- Different programs and criteria that promote FAW will be defined and controlled. The latter also implies that the payments necessary to compensate farmers for taking part in the various programs needs to be agreed on.12
- Implementing any of those higher FAW standards at the farm level remains voluntary. However, those producers who put defined FAW measures into practice are rewarded by a payment that is independent of the price for pigs at the farm level.
- Prices for pig meat at the consumer level are as well not differentiated according to the different animal welfare standards in the program.
- Communication of the scheme vis-a-vis the consumer and the public at large will be done in a comprehensive manner and as a joint effort. Advertisement at the product level is not envisaged.
- Implementation and control is adjoined to the existing quality assurance scheme QS which already covers about 95% of German pig production thereby easing co-ordination and limiting transaction costs (QS, 2014).

---

11 The initiative for poultry is organised slightly different given the different market structure in this sector (see QS, 2013).
12 This is, as discussed above, no trivial task.

Farm Animal Welfare: A Challenge for Markets and Policy 55

This industry self-regulation strategy provides an economic incentive for producers to accept higher animal welfare standards. Thus, it considers the trade-off they face between productivity and FAW standards. At the same time it solves the free rider problem at the level of consumer though not at the citizen level.

The amount of money retailers participating in the initiative pay into the fund depends on the quantity of QS labelled meat they sell. This implies an increase of per unit costs at that level. It can be assumed that retailers will pass on the additional cost for meat products onto their customers, likely as a mark-up on meat products. The latter would imply that only meat consumers cover the costs and that e.g. vegans, though they might benefit from the initiative do not contribute in financing the system. Given that the additional per unit costs are the same for all retailers participating in the scheme, the effect on their relative competitive position is assumed to be limited.[13]

Though the free rider problem at the level of consumers is solved by such a scheme, it still exists at the retail level. Those retailers not taking part in such a co-ordinated system could gain a competitive advantage which is an incentive not to join or to leave the initiative. Thus the system is only viable if (almost) all members of the meat value chain participate. Regarding the "Initiative Animal Welfare" all relevant retailers in Germany have agreed to take part in the system so that the market coverage is high (see QS, 2013). Thus, though some smaller retail companies may have potential competitive advantages, this does not carry much weight. In addition, for those having agreed to and joined the initiative it might be a considerable loss of reputation if they would quit and not stick to their commitment as this likely would receive considerable media attention.

The "Initiative Animal Welfare" has not started yet. Even though, the main principle of this scheme is simple, there is a lot still to agree upon before it can be implemented. E.g. in spite of all the difficulties in defining animal welfare, the system requires appropriately determined measures for increasing animal welfare as well as a scheme for rewarding producers adequately. In addition, open questions remain with respect to its effects on meat products sold under already existing animal welfare labels. Though these labels still can be applied to differentiate products with standards exceeding those of the "Initiative Animal Welfare", the initiative might have a negative impact on the competitiveness of other labels as it reduces the gap between those products and the mass market. Thus, it could counteract animal welfare initiatives of other private actors. Finally, co-operation

---

13 However, the relative change in costs and thus in the mark up for consumers might well differ, e.g. between discounters and other retail stores, and thus might impact the respective competitive position following the introduction of the scheme. On the other hand, one could argue that especially for low price retail chains participation in the scheme could be attractive as it has the potential to raise their overall reputation.

among competitors implies the danger of exercising market power and thus needs to be agreed upon by Germany's federal anti-trust agency (Bundeskartellamt). Also this is still to be achieved.

Conclusions

The public discussion about animal welfare can be taken as an indicator for poor market performance regarding the implementation of a societal desired way of animal husbandry. From an economic point of view, interventions by the government can be justified under the following circumstances; asymmetric information prevails, externalities exist or FAW is a public or a merit good.

In case of market failure as a consequence of asymmetric information, government interventions like education programs or the provision of information, tailored to the different requests and needs of consumers might increase consumers' knowledge with respect to FAW issues, their awareness, the visibility of the characteristic at the point of sale and thus, potentially its relevance in consumers' purchase decision. As has been discussed in this paper the EU is contemplating of introducing measures such as the provision of standardised farm animal welfare indicators on a mandatory or voluntary basis to support interested consumers in identifying and choosing welfare-friendly products.

Assuming that FAW is a public or a merit good can provide additional arguments for state intervention in markets, e.g. by increasing FAW standards in order to close the free-rider gap or the consumer-citizen gap, respectively. However, such an intervention is only desirable if it has a positive impact on overall societal welfare. Higher FAW standards affect production costs and thereby competitiveness. Divergent standards between countries may result in trade flows in favour of those countries with lower standards, thus harming producers in countries with stricter FAW regulations and limiting the potential positive effects for the welfare of farm animals. This holds, especially, as governmental action to prevent or limit imports of lower standard meat products and/or to subsidy high animal welfare livestock products are restricted by WTO commitments.

Self-regulating strategies of the meat sector can help to overcome (part of) the consumer - citizen gap by rising private standards for animal welfare if it is co-ordinated. In order to succeed members of the value chain have to prevent a free-rider problem at the retail level and to commonly agree on the different FAW measures and how to support their implementation. The German 'Initiative Animal Welfare' is an interesting example in this respect. Even though the principles of such an approach are straightforward, the details are tricky. In addition, such a strategy needs to be accepted by Germany's federal anti-trust agency.

Thus, the exposition reveals that the issues linked to farm animal welfare are far from trivial and their improvement need the effort of all members of the meat value chain as well as the government. Thus, in line with the EU COMMISSION we like to conclude that "everyone is responsible" (DIRECTORATE-GENERAL FOR HEALTH AND CONSUMERS, 2010). Better FAW can only be realized if this is acknowledged.

List of References

Alphonce, R., Alfnes, F. and Sharma, A. (2013): Voting or Buying: Inconsistency in Preferences toward Food Safety in Restaurants. Paper presented at the Joint AAEA & CAES Annual Meeting, Washington, DC, August, 4-6.

Besley, T. (1988): A simple model for merit good arguments. Journal of Public Economics, 35, 371-383.

Bock, B.B. and van Huik, M.M. (2007): Animal welfare: the attitudes and behavior of European pig farmers, British Food Journal, 11, 931-944.

Camerer, C., Issacharoff, S., Loewenstein, G., O'Donoghue, T. and Rabin, M. (2003): Regulations for Conservatives: Behavioral Economics and the case for "asymmetric paternalism"

Croney, C.C. and Millman, S.T. (2007): Board-Invited Review: The ethical and behavioral bases for farm animal welfare legislation. J. Anim. Sci., 85, 556–565.†

D'Amico, D. (2009): Merit goods, paternalism, and responsibility. Paper presented at the XXI Conference on Public Choice and Political Economy. Pavia University. 24–25 September 2009.

Directorate-General for Health and Consumers (2010): Everybody is responsible. Animal Welfare Newsletter. Brussels.

Duncan, I.J.H. and Dawkins, M.S. (1983): The problem of assessing 'well-being' and 'suffering' in farm animals, In: Smidt, D. (ed.), Indicators Relevant to Farm Animal Welfare, Martinus Nijhoff, The Hague, 13–24.

Duncan, I.J.H. (2005): Science-based assessment of animal welfare: farm animals, Revue Scientifique et Technique (International Office of Epizootics), 24, 483-492.

Edwards, J.D. (2004): The role of the veterinarian in animal welfare: a global perspective, In: Proc. Global Conference on animal welfare: an OIE initiative, Paris, 27-35.

Estevez, I. (2003): Animal Welfare in Modern Animal Agriculture. In: Reynnells, R.: The Science and Ethics Behind Animal Well-Being Assessment One in a Series of Educational Programs Presented by the Future Trends in Animal Agriculture, Washington D.C., 6-11.

EU Commission (2007): Eurobarometer: Attitudes of EU Citizens towards Animal Welfare, Special Eurobarometer, 270/Wave 66.1, Brussels.

EU Commission (2009): Report from the Commission to the European Parliament. the Council, the European Economic and Social Committee and the Committee of the Regions: Options for animal welfare labelling and the establishment of a European Network of Reference Centres for the protection and welfare of animals, Brussels.

Fraser, D., Weary, D.M., Pajor, E.A. and Milligan, B.N. (1997): A scientific conception of animal welfare that reflects ethical concerns. Animal welfare, 6, 187-205.

Grunert, K.G. (2006): Future trends and consumer lifestyles with regard to meat consumption. Meat Science, 74, 149-160.

Hartmann, M. (2011): Corporate social responsibility in the food sector. European Review of Agricultural Economics, 38 (3), 297-324.

Harvey, D. and Hubbard, C. (2013): Reconsidering the political economy of farm animal welfare: An anatomy of market failure. Food Policy, 38, 105-114.

Hewson, CJ. (2003): What is animal welfare? Common definitions and their practical consequences. Can Vet J, 44, 496-499.

Ingenbleek, P.T.M., Immink, V.M., Spoolder, H.A.M. and Bokma, M.H. (2013): EU animal welfare policy: Developing a comprehensive policy framework. Food Policy, 37, 690-699.

Kirchgässner, G. (2012): Sanfter Paternalismus, meritorische Güter, und der normative Individualismus. School of Economics and Political Science, Department of Economics, University of St. Gallen, Discussion Paper no. 2012-17.

Lagerkvist, C.J. and Hess, S. (2011): A meta-analysis of consumer willingness to pay for farm animal welfare. European Review of Agricultural Economics, 38 (1), 55-78.

Lassen, J., Sandøe, P. and Forkman B. (2006): Happy pigs are dirty! – conflicting perspectives on animal welfare. Livestock Science, 103, 221-230.

McInerney, J. (2004): Animal Welfare, Economics and Policy. London, UK.

Mann, S. (2003): Why organic food in Germany is a merit good. Food Policy, 28, 459–469.

Mann, S. and Gairing, M. (2012): Does libertarian paternalism reconcile merit goods theory with mainstream economics? Forum for Social Economics, XLI (2-3), 206-219.

Medema, S.G. (2014): The Curious Treatment of the Coase Theorem in the Environmental Economics Literature, 1960–1979. Review of Environmental Economics and Policy, 8 (1), 39-57.

Müller, M. and Schmitz, P.M. (2002): Ökonomische, ethische und medizinische Relevanz zur Beurteilung ausgewählter Tierhaltungsverfahren und -systeme auf der Basis der Conjoint-Analyse. Schriftenreihe der Landwirtschaftlichen Rentenbank, 17, 7-47.

Musgrave, R.A. (1956): A multiple theory of budget determination. Finanzarchiv, XVII (3), 333-343.

Napolitano, F., Girolami, A. and Braghieri, A. (2010): Consumer liking and willingness to pay for high animal welfare products. Trends in Food Science and Technology, 21, 537-543.

Norwood, B.F. and Lusk, J.L. (2011): Compassion by the Pound: The Economics of Farm Animal Welfare. Oxford University Press, New York.

Olsen, M. (2000): Power and Prosperity – Outgrowing Communist and Capitalist Dictatorships. Basic Books, New York.

Olynk, N.J. (2012). Assessing changing consumer preferences for livestock production processes. Animal Frontiers, 2 (3), 32-38.

QS (2013): Initiativen zum Tierwohl. Absichtserklärung zur Umsetzung der Initiativen zum Tierwohl für Schwein und Geflügel. Berlin, Retrieved on 24.3.2014 from http://www.q-s.de/initiative_zum_tierwohl_1.html.

QS (2014): Ihr Prüfsystem für Lebensmittel. Retrieved on 24.3.2014 from http://www.q-s.de/

Spinka M. (2006): How important is natural behaviour in animal farming systems? Appl. Anim. Behav. Sci., 100.117-128.

Farm Animal Welfare: A Challenge for Markets and Policy 59

Swanson, J.C., Lee, Y., Thompson, P.B., Bawden, R. and Mench, J.A. (2011): Integration - Valuing stakeholder input in setting priorities for socially sustainable egg production. Poultry Science, 90, 2110–2121.
Swanson, J. C. (2008): The ethical aspects of regulating production. Poultry Science, 87, 373-379.
Sunstain, C.R. and Thaler, R.H. (2003): Libertarian Paternalism Is Not an Oxymoron, University of Chicago Law Review, 70. 1159-1202.
Te Velde, H., Aarts, N. and van Woerkum, C. (2002): Dealing with ambivalence: farmer's and consumer's perceptions of animal welfare in livestock breeding, Journal of Agricultural and Environmental Ethics, 15, 203-219.
Toma, L., McVittie, A., Hubbard, C. and Stott, A. (2011): A Structural equation model of the factors influencing British consumers' behaviour toward animal welfare. Journal of Food Products Marketing, 17, 261-278.
Tucker, C.B., Waery, D.M. and D. Fraser (2003): Effects of Three Types of Free-Stall Surfaces ob Preferences and Stall usage by Dairy Cows. American Dairy Science Association, 86, 521-529.
Vanhonacker, F., Verbeke, W., van Poucke, E., and Tuyttens, F.A.M. (2007): Segmentation based on consumers' perceived importance and attitude toward farm animal welfare, International Journal of Sociology of Food and Agriculture, 15 (3), 84-00.
Vanhonacker, F., Verbeke, W., Van Poucke, E. and Tuyttens, F.A.M. (2008): Do citizens and farmers interpret the concept of farm animal welfare differently? Livestock Science, 116, 126-136.
Verbeke, W. (2005): Agriculture and the food industry in the information age. European Review of Agricultural Economics, 32, 347-368.
Ver Eecke, W. (1998): The concept of a 'merit good': the ethical dimension in Economic Theory and the history of economic thought of the transformation of economics into socio-economics. Journal of Socio-Economics, 27 (1), 133-153.
Wilkie, R. (2005): Sentient commodities and productive paradoxes: The ambiguous nature of human-livestock relations in Northeast Scotland. Journal of Rural Studies, 21 (2), 213-230.
WTO (1995a): Agreement on Agriculture. Retrieved on 28.03.2014. http://www.wto.org/english/docs_e/legal_e/14-ag.pdf.
WTO (1995b): Agreement on the Application of Sanitary and Phytosanitary Measures, Article 5. Retrieved on 28.3.2014. http://www.wto.org/english/docs_e/legal_e/15-sps.pdf.
WTO (2013a): WTO Ministerial Conference, Ninth Session, Bali, 3-6 December, Declaration and Decisions. Retrieved on 28.3.2014 from http://wto.org/english/thewto_e/minist_e/mc9_e/bali_texts_combined_e.pdf.
WTO (2013b): Goods schedules – Current situation of schedules of WTO members. Retrieved on 28.03.2014 from http://www.wto.org/english/tratop_e/schedules_e/goods_schedules_table_e.htm.

From Policy Analysis to Recommendations for Evidence-based Food Policy: Some Thoughts on „New" Policy Instruments

Roland HERRMANN, Rebecca SCHRÖCK and Matthias STAUDIGEL

Abstract

In agricultural economics, researchers have contributed over decades to the policy dialogue in very different ways. They provided theoretical and empirical policy analyses, they made policy recommendations in scientific advisory groups and sometimes they were engaged in their role as public economists to defend the public interest. Recently, researchers have been challenged by the request developed in economics and in administration to elaborate evidence-based analysis and advice. According to the concept, it should be analysed, whether the instruments fulfil the objectives defined by politicians and, if yes, at lowest costs. Impact analyses on the effectiveness and efficiency of policies ought to be performed with the econometric approach to program evaluation based on natural experiments. We show that this trend is a major challenge in the case of "new" policy instruments targeted at food quality or health. The protection of geographically differentiated foods as well as food taxes are not yet evidence-based in many ways. We argue that evidence-based policy analysis is to become a most important part of the toolbox, but agricultural economists should continue to go beyond it by adding questions, objectives and instruments that are not predefined by policymakers.

Introduction

There is a consensus that research can support economic policy with regard to the choice, implementation or correction of policy instruments. The way in which researchers express their views of policy, however, may take very different forms: In academic work addressed to their peers, researchers may analyse the implications of policy instruments in *theoretical policy analyses*. In doing so, policy options can be defined freely and may go beyond the instruments politicians are actually considering. Researchers may also combine theoretical and quantitative methods in their *empirical policy analyses*. Theoretical and empirical policy studies can either be based on positive analysis if the situations with and without policy are compared or on normative analysis if an optimal policy is derived from an optimization approach. Quite strong conclusions can be drawn from such studies on the costs and benefits of policy instruments. Positive as well as normative analyses will include value judgements on the policy instruments analysed and the methodology but can and should be free from value judgements in the analysis itself.

Researchers go further when they make *policy recommendations*. The researcher has to reveal his own objective function and to critically verify the model assumptions and coefficients on which the recommendations are based, even if uncertainty remains. In that sense, more judgements are needed but policy recommendations may still be based on scientific policy analysis to a very large extent. Researchers may be asked to provide policy recommendations as members of scientific advisory groups to ministries, organizations or politicians. Sometimes, they will do it voluntarily in order to influence the public debate or even political decisions in a direction they regard as desirable.

Some economists go even further when they see their role as *public economists*. GIERSCH (1990, p. 16) defined the public economist as an economist who represents the public interest in policy debates. Being a public economist would, e.g., include to discover inefficiencies in the context of governmental interventions and to "stimulate the public policy debate by stating the case for or against proposals under discussion or by suggesting superior solutions". The public economist will regard it as his duty to inform the society on implications of current or future policy irrespective of whether politicians ask for his opinion. Typically, the public economist will (i) often express his views in the media and (ii) find it difficult or impossible to distinguish strictly between policy analysis, policy recommendations and value judgements.

Michael SCHMITZ was involved in all these fields in his academic work on agricultural and food policy as only some examples clearly show. In a theoretical policy analysis, he elaborated that important agricultural trade restrictions will affect the instability of world agricultural markets very differently (SCHMITZ 1984). In his studies on the sugar market, he combined theory and quantitative techniques in empirical policy analyses to identify impacts of the European Sugar Policy on the European Community (HERRMANN and SCHMITZ, 1984) and on developing countries (KOESTER and SCHMITZ, 1982). As a member of the Scientific Advisory Council of the German Federal Ministry of Food, Agriculture, and Consumer Protection, he contributed to many policy recommendations and in some cases, such as the advisory opinion on how the government should cope with yield and price risks in agriculture, he was in charge of the final report (WISSENSCHAFTLICHER BEIRAT, 2011). And in his role as a public economist, he criticized the closure of the Central Marketing Agency in Germany and designed a follow-up solution (SCHMITZ and HESSE, 2009).

New challenges for policy analysis and recommendations arise from the credibility problem the economic profession is facing after the Financial Crisis (ANGRIST and PISCHKE, 2010) and in the context of new methodological tools (IMBENS and WOOLDRIDGE, 2009). The need to rely on evidence-based policymaking and advice has been stressed for economic policy (WISSENSCHAFTLICHER BEIRAT, 2013) as well as for consumer policy (WISSENSCHAFTLICHER BEIRAT VERBRAUCHER- UND ERNÄHRUNGSPOLITIK, 2013). Researchers are expected to provide evidence-based policy recommendations to politicians and the society. Advice should rely on instruments for which it is empirically proven that they are effective and efficient, i.e. they realize their objectives at lowest costs.

In the following, we will focus on evidence-based policy advice. We will address the question whether highly debated "new" instruments of food policy are actually evidence-based: (i) protected geographical indications for high-quality foods; (ii) taxes on unhealthy foods as a countermeasure against the obesity problem.

The chapter is organized as follows. After this Introduction, the main strands of food policy analysis in the past are analysed briefly and the concept of an evidence-based food policy is clarified. Following, it is investigated both for the regulation of geographically differentiated foods and for food taxes as a health-policy instrument to which extent the conditions for an evidence-based policy are fulfilled. We analyse whether empirical proofs do exist that geographical indications and food taxes fulfil their stated objectives efficiently. Finally, findings are summarized and conclusions are drawn on whether the "new" food policy instruments can be justified as being evidence-based policies and which types of analyses are lacking in order to show the effectiveness or efficiency of these instruments.

Policy Analysis: A Review of the Literature

Analysis of Traditional Agricultural and Food Policy

For a rather long time, agricultural and food policy in industrialized countries has been dominated by agricultural price policy. Therefore, a comprehensive toolbox for the analysis of impacts arising from such agricultural market policies has been developed. Based on welfare economics (JUST et al, 2004) and single-market and multimarket models for commodities, the allocative and redistributive implications of alternative agricultural and food price policies were elaborated (ALSTON and JAMES, 2002; GARDNER, 1987; SUMNER et al., 2010). It was typical for this literature to model the effects of changing price levels due to agricultural and food policies on consumers, producers, the government and the society as a whole.

Extensions of mainstream agricultural policy analysis included the economics of commodity price stabilization (NEWBERY and STIGLITZ, 1981), agricultural trade policy (HOUCK, 1986), food security (BIGMAN, 1982), agricultural research policy (ALSTON et al., 1995), the economics of food safety (ANTLE, 2001) as well as the political economy of agricultural and food policy (RAUSSER, et al., 2011).

Early and important econometric contributions of agricultural economists to the methodology of applied econometrics are surveyed by FOX (1986). In general, agricultural economics has a "long and rich history of fundamental contributions to the literature on econometric and statistical methods and economic measurement" (BESSLER et al., 2010, p. 571). Consequently, it is not surprising that agricultural market analyses, including many supply and demand studies, have a strong quantitative tradition (MYERS et al., 2010). Moreover, policy analyses have for a long time been based on econometrically estimated coefficients such as price elasticities of supply and demand and price transmission elasticities. Cases in point are comprehensive ex-ante studies on the implications of international agricultural trade liberalization on prices, trade and welfare in developed and developing countries (TYERS and ANDERSON, 1992), ex-post studies on the impacts of developing countries' agricultural and macroeconomic policies on developed and developing countries (SCHIFF and VALDÉS, 1992; ANDERSON, 2010; LLOYD et al., 2010) or ex-ante and ex-post studies on the economics of commodity promotion (KAISER et al., 2005).

Evidence-based Policy Analysis and Recommendations

The recent discussion on evidence-based economic policy has its roots in medical science. Evidence-based medicine has been the major direction of best-practice medical care for many years. SACKETT et al. (1996, p. 71), in their often-cited Editorial, define: "Evidence-based medicine is the conscientious, explicit and judicious use of current best evidence in making decisions about the care of individual patients. The practice of evidence based medicine means integrating individual clinical expertise with the best available external clinical evidence from systematic research." Important is the empirical proof that a treatment or a medicine has been effective.

In evidence-based economic policy, a similar empirical proof is crucial. A study by the Scientific Advisory Council of the German Federal Ministry for Economic Affairs and Energy (WISSENSCHAFTLICHER BEIRAT, 2013) is typical for the recent trend towards evidence-based policy recommendations. The objectives of economic policy evaluation are elaborated and evaluation is defined as impact analysis. According to the authors, two major questions shall be answered:

- Does the government reach its objectives with the policy instrument chosen and to what extent?
- Could the government reach the same impact at lower costs?

The first question refers to *effectiveness*, the second one to *efficiency*.

Although these two particular questions have been addressed before in agricultural and food policy and sound familiar, the authors of the report go beyond the approaches discussed so far. They link the question of effectiveness and efficiency of policies to a new development in empirical economics, i.e. the new experimental approach in the econometrics of program evaluation (WISSENSCHAFTLICHER BEIRAT, 2013, pp. 10-17). Basically, a strong case is made for the use of random natural experiments in program evaluation. Like in medicine, treatment and control groups shall be defined in well-designed experiments in order to allow an unambiguous identification of policy impacts. For instruments applied in labour market, education or industrial policy, one would want to compare on the basis of microdata whether subjects under the influence of a program would perform significantly better than their counterparts which did not participate in the program.

Even if a controlled randomization is not possible, the Scientific Advisory Council (ibid., p. 13) argues that quasi-experimental approaches such as the difference-in-difference approach, the regression-discontinuity method or instrumental-variable techniques may be utilized in order to come close to a randomized trial. The rationale for program evaluation on the basis of the experimental approach is presented in the surveys by IMBENS and WOOLDRIDGE (2009) and in ANGRIST and PISCHKE (2010). The econometric methodology is comprehensively covered by LEE (2005).

Evidence-based approaches aim at the improvement and credibility of policy analyses and recommendations (JENSEN, 2013). In that sense, they represent a major challenge for food policy analyses and recommendations, too. Policy advice on agricultural and food policy instruments is often not based on *empirical* impact analyses. This holds even more for "new" instruments of food policy that are targeted at food safety or health issues rather than agricultural market support. Here, an evidence-based policy analysis and advice seems very necessary. Empirical challenges will be discussed in detail in the following two sections for the regulation of geographically differentiated products and in for food taxes as health-policy instruments. Questions remain on the choice of methodology in empirical policy analysis and we will answer those in the final conclusions.

The Regulation of Geographically Differentiated Products: Thoughts on Effectiveness and Efficiency

With the liberalization of international agricultural markets, policy instruments have markedly changed. European agricultural and food policy has moved from agricultural market and price support, which had dominated over decades, to new policy instruments which are targeted towards food safety and the quality of foods. In the European quality policy, the protection of geographical indications (GIs) of foods has become a major instrument (BECKER, 2009) for high-quality foods. Products with high collective reputation and a strong link between origin and product quality such as Chianti, Champagne or Parma ham are cases in point. Under that system, product names with a geographical origin can only be used by producers of the particular region who stick to specified rules of production for the registered product. Two types of protected names are distinguished, i.e. the protected designation of origin (PDO) and the protected geographical indication (PGI). The PDO covers agricultural products and foodstuffs which are produced, processed *and* prepared in a given geographical area. For a PGI, *at least one* of these stages has to take place in the area. These regulations coincide with an increasing role of the geographical origin and regional products in the valuation of foods by consumers (ITTERSUM et al., 2007).

Figure 1 displays the rationale as well as expected economic effects of geographically differentiated food products. An often-cited economic rationale for the protection of GI products is a potential market failure along the lines of AKERLOF's lemon case which is due to information asymmetry and quality uncertainty (AKERLOF, 1970). On unregulated markets, a high-quality product with proven regional origin would be susceptible to counterfeiting. Imitation would lead to a coexistence of higher- and lower-quality foods and consumers would suffer from incomplete information and quality uncertainty. Prices may be depressed as a consequence of the low-quality supply. Incentives to offer high-quality foods might then be limited and, if screening by consumers and signalling by producers do not work, a market failure will occur if high quality products are displaced from the market.

Vice versa, if a product's high quality is due to the production technology and the knowledge in a given geographical area, i.e. the terroir, the protection of the GI product may avoid a market failure in the sense of AKERLOF. Quality assurance and labelling of the high-quality product may lead to an increasing demand as quality uncertainty is reduced. Furthermore producers may face additional costs for certification, labelling and advertising as we know from quality assurance schemes (WILL, 2013). Prices may increase, producer gains will occur and consumers may benefit from better information on the market. Rural income may be raised as well as overall welfare. Improvements in environmental protection may be a further economically valuable long-term impact.

Major elements of this economic rationale are laid down in Regulation No. 1151/2012 (EUROPEAN COMMISSION, 2012) which is the legal basis of the European quality schemes for agricultural products and foodstuffs. In the §§ 1 to 6, the background and objectives of quality policy are formulated: (i) Consumers "increasingly demand quality as well as traditional products" and value "the diversity of the agricultural production in the Union" and, thus, appreciate agricultural products and foodstuffs "linked to their geographical origin" (§2). (ii) Producers will offer "a diverse range of quality products if they are rewarded fairly for their effort" (§3) and this requires that products can be identified correctly on the market. (iii) These incentives can benefit the rural economy (§4) by their income-raising effects and will contribute to economic and social cohesion (§5).

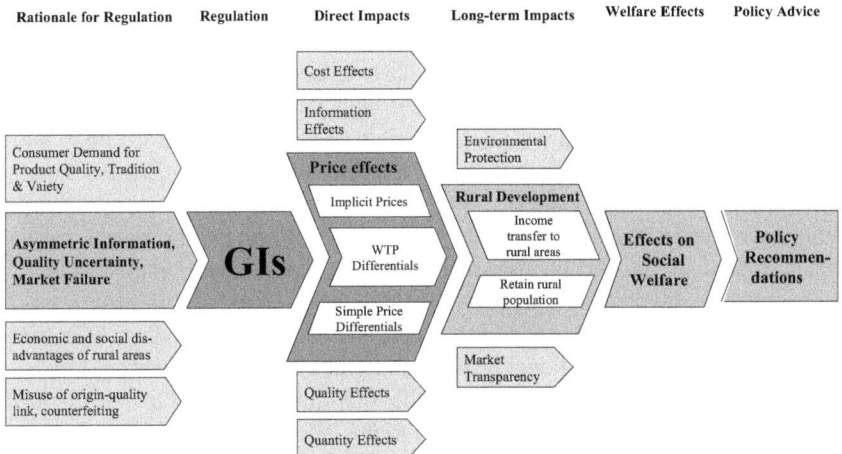

Fig. 1: *Rationale and Expected Effects of Geographical Indications and Regional Food Products*

Source: Own presentation

In order to assess the key areas of research, we have reviewed 28 studies analysing effects of GIs and/or country of origin[1]. Comprehensive surveys covering several stages and considering various indicators of efficiency and effectiveness of GIs are provided by HERRMANN and TEUBER (2011) and the EUROPEAN COMMISSION (2008).

Most of the existing studies focus on only one single aspect of the causal chain, i.e. on price impacts. 25 out of the 28 studies in our meta-analysis investigate price effects. Among them, twelve studies use hedonic approaches and give implicit prices for GI labels or distinct product origins, eleven studies apply contingent valuation and discrete choice techniques and indicate willingness-to-pay

---

1 Detailed information on the evaluation scheme and the results of the meta-analysis of studies analyzing effects of geographically differentiated food products is available from the authors upon request.

measures. Two studies simply compare the prices of geographically differentiated and conventional products. DESELNICU et al. (2013) provide a meta-analysis of studies estimating price premiums for geographically differentiated food products. Direct impacts on producer costs (5 studies), quantities sold (6), product quality (5) and consumer information status (3) are rarely studied. Only a few studies such as LENCE et al. (2007), MOSCHINI et al. (2008) and ZAGO and PICK (2004) derive effects on social welfare within theoretical models.

The results of the meta-analysis underline both that the price impact is a key variable in empirical analyses and that the price premium consumers are willing to pay for geographically differentiated foods is an indicator often applied. It is an ambiguous question whether the latter leads to an additional producer incentive and, thus, an income gain for producers and higher rural incomes. The answer crucially depends on the relative size of the rightward shift of demand due to the introduction of a PDO or PGI compared to the leftward shift of the supply function. Figure 2 shows this argument.

Suppose that a higher-quality product from a specific geographical area exists but cannot be distinguished by consumers from the lower-quality mass product. On a competitive market, both products would then be offered at a uniform mass-market price p_0. At that price, S is the supply of the regional product, D is the depressed demand for the regional product under incomplete information.

If a PDO or a PGI is now introduced, the high-quality product can be distinguished from the lower-quality product due to the label and informative advertising. Quality uncertainty is reduced and consumers realize a higher demand level, i.e. D'. Additional costs arise by participating in the producer club due to additional control costs and costs of advertising. The supply curve shifts from S to S'. On a competitive market, consumers will now pay a price p'_C that is above the mass-market price p_0 in the new equilibrium. Regional producers will have to pay additional costs by participating in the PDO or PGI scheme; their price p'_P is below the consumer price p'_C.

It is important here that demand has shifted more than supply. Therefore, the introduction of the PDO or PGI has raised the net producer price from p_0 to p'_P. This is the precondition for a producer welfare gain as can be seen from an alternative and lower demand shift from D to D''. With the supply curve S' and the demand curve D'', the PDO or PGI would raise the consumer price from the mass-market price p_0 to p''_C and lower the producer price to p''_P. Thus, it is possible that a PGO or PDI leads to a price premium which consumers pay ($p''_C - p_0 > 0$) and a price discount which producers experience ($p''_P - p_0 < 0$) at the same time. Figure 2 illustrates that the existence of a price premium paid by consumers does not necessarily imply a welfare gain due to a PDO or PGI for the producer club. Only if labelling and advertising shifts the demand curve more than the supply curve, producers will get a price premium, too, and this will also be associated with higher purchases ($q'-q_0$) and a rising market share of the GI product.

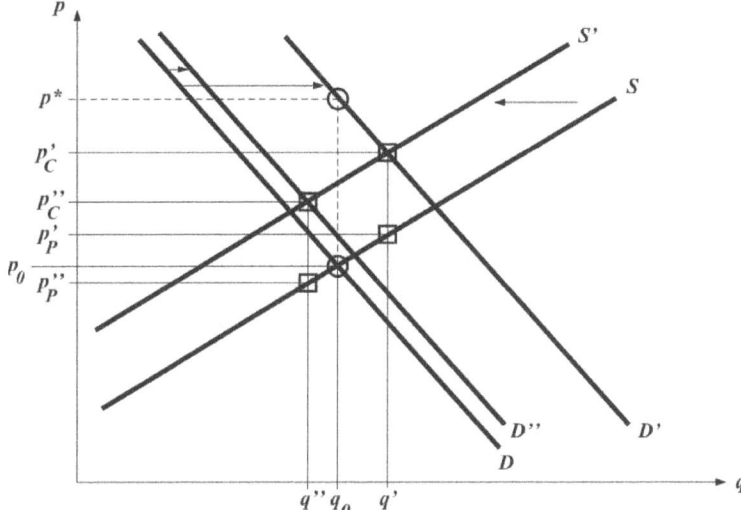

Fig. 2:   *The Price Premium and the Effectiveness of Protected Geographical Indications for Foods*

Source: Own presentation

Suppose that a higher-quality product from a specific geographical area exists but cannot be distinguished by consumers from the lower-quality mass product. On a competitive market, both products would then be offered at a uniform mass-market price p_0. At that price, S is the supply of the regional product, D is the depressed demand for the regional product under incomplete information.

If a PDO or a PGI is now introduced, the high-quality product can be distinguished from the lower-quality product due to the label and informative advertising. Quality uncertainty is reduced and consumers realize a higher demand level, i.e. D'. Additional costs arise by participating in the producer club due to additional control costs and costs of advertising. The supply curve shifts from S to S'. On a competitive market, consumers will now pay a price p'_C that is above the mass-market price p_0 in the new equilibrium. Regional producers will have to pay additional costs by participating in the PDO or PGI scheme; their price p'_P is below the consumer price p'_C.

It is important here that demand has shifted more than supply. Therefore, the introduction of the PDO or PGI has raised the net producer price from p_0 to p'_P.

This is the precondition for a producer welfare gain as can be seen from an alternative and lower demand shift from D to D''. With the supply curve S' and the demand curve D'', the PDO or PGI would raise the consumer price from the mass-market price p_0 to p_C'' and lower the producer price to p_P''. Thus, it is possible that a PGO or PDI leads to a price premium which consumers pay ($p_C'' - p_0 > 0$) and a price discount which producers experience ($p_P'' - p_0 < 0$) at the same time. Figure 2 illustrates that the existence of a price premium paid by consumers does not necessarily imply a welfare gain due to a PDO or PGI for the producer club. Only if labelling and advertising shifts the demand curve more than the supply curve, producers will get a price premium, too, and this will also be associated with higher purchases ($q' - q_0$) and a rising market share of the GI product.

Unfortunately, the information needed to quantify the situations with and without policy in Figure 2 is usually not provided in empirical studies. Most empirical price-premium studies do not indicate whether producers gain or lose from GI differentiation. As illustrated in Figure 1, price impacts of geographical indications have been measured in the literature in three alternative ways: (i) as the marginal willingness to pay for the characteristic "geographical origin"; (ii) as implicit prices for the same characteristic in hedonic analyses; (iii) as the simple price difference between the observed price of the GI product and any price of an unlabelled benchmark product.

First, if the additional willingness to pay for the characteristics "geographical origin" is measured by conjoint or discrete-choice analyses, this is a purely consumer-oriented measure of the price premium. In Figure 2, the additional willingness to pay for the characteristic "PDO" or "PGI" can be illustrated as ($p^* - p_0$). As the supply side of GI labelling is ignored and a totally inelastic supply is implicitly assumed, this procedure will overestimate the price premium consumers are actually paying for the difference between the observed price of the GI product and any price of an unlabelled benchmark product. Moreover, impacts on producer prices are not captured and, therefore, the approach is not helpful in measuring producer incentives arising from PGI or PDO introduction.

Secondly, if hedonic analysis is used and follows the supply-and-demand framework for product characteristics by ROSEN (1974), it is possible to quantify the price premium for GI products outlined in Figure 2. The price difference ($p_C' - p_0$) could then be derived from a hedonic model as a measure of the implicit price of the labelled geographical origin of a food product. However, it is not possible to derive the (potential) price premium for producers directly from the hedonic results as producer incentives under the influence of the label are typically unobserved. It is necessary to quantify the marginal costs arising from participation in the producer club additionally, as indicated by ($p_C' - p_P'$) in Figure 2.

It would then be possible to calculate the GI-induced producer premium ($p'_p - p_0$) in an additional step following the hedonic analysis.

Thirdly, several studies approximated the price premium for GIs by comparing the price of the geographically differentiated product with the price of a substitutive product characterizing the mass market. This procedure is, differently from the former two methods, not theory-driven and seems ad hoc. The price premium derived may or may not be close to the premium which arises from a model-based approach comparing the situations with and without policy.

Starting from Figure 2, we can summarize that GI-induced price premia which consumers pay and producers receive can be computed but have been treated superficially up to now. One option is a hedonic supply-and-demand-model complemented by a quantitative analysis of the marginal costs of participation in the PDO or PGI scheme. Alternatively, a structural market model could be estimated econometrically and used for a computation of the premium consumers are willing to pay. Again complemented by an analysis of participation costs, it would also be possible to calculate the premium realized by producers of the geographically differentiated food. These empirical approaches could then provide the basis for modelling the consequential impacts of GI protection on supply, demand, earnings, costs, income and economic welfare.

A stylized model of welfare implications of GI protection is presented in Figure 3. Analogous to Figure 2, we assume that GI protection will improve market information and reduce quality uncertainty. Differential qualities of the higher- and lower-quality products are now visible for consumers. The demand curve shifts to the right as a consequence of GI labelling and informative advertising, i.e. from D to D'. Participation of producers in the PDO or PGI raises marginal costs and shifts supply from S to S'. As the demand shift is stronger, the net price of producers rises (from p_0 to p'_p) as does the price consumers have to pay for the geographically differentiated product (from p_0 to p'_c).

A stylized model of welfare implications of GI protection is presented in Figure 3. Analogous to Figure 2, we assume that GI protection will improve market information and reduce quality uncertainty. Differential qualities of the higher- and lower-quality products are now visible for consumers. The demand curve shifts to the right as a consequence of GI labelling and informative advertising, i.e. from D to D'. Participation of producers in the PDO or PGI raises marginal costs and shifts supply from S to S'. As the demand shift is stronger, the net price of producers rises (from p_0 to p'_p) as does the price consumers have to pay for the geographically differentiated product (from p_0 to p'_c).

The increasing net price raises producer welfare by the area (a + b + c). The welfare implications of GI labelling and informative advertising for consumers depend crucially on the assumptions regarding (i) whether the information provided is correct or wrong; (ii) whether the initial information of consumers on quality was correct or false; (iii) whether preferences change due to labelling and advertising (JUST et al., 2004, Chapter 11). We posit here that (i) GI labelling and informative advertising shifts consumer demand to a level with correct and full quality information, i.e. D'; (ii) consumers would be on the "wrong" demand curve D if they suffered from incomplete quality information in the situation without GI regulation; (iii) consumer preferences will not change. Under these assumptions, consumer surplus will be (a + b + d + e + f) in the initial situation without GI regulation and (f + g) afterwards. As search costs are reduced, consumers gain area g due to additional quality information at the new price p'_c. They lose (a + b + d + e) due to the price increase GI regulation will cause. A consumer gain will arise if g > (a + b + d + e). The overall welfare effect will be (c + g - d - e) which will be positive under the assumptions made.

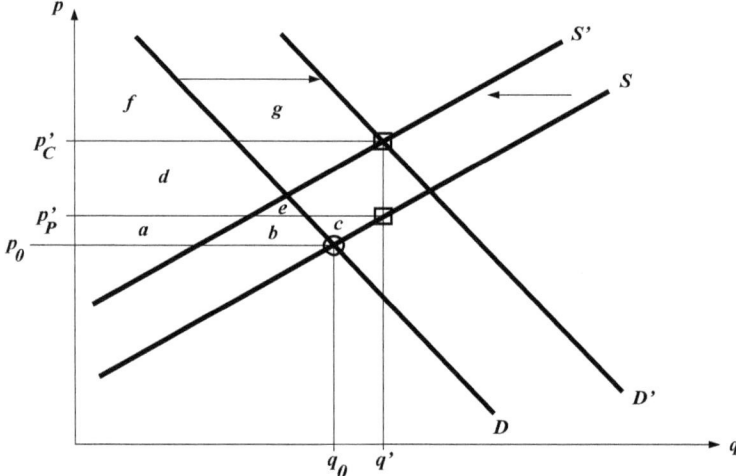

Fig. 3:   Labelling of and Informative Advertising for a Geographical Indication: Welfare Implications

Source: Own presentation

It has been discussed in the literature (HERRMANN and TEUBER, 2011, pp. 818 f.) that the assumptions made are strong. Therefore, substantial doubt remains whether the overall welfare effect of GI regulation is actually positive and how welfare effects are distributed. Most likely, this question will have to be clarified on the basis of an empirical case-by-case basis. There are no studies which quantify these welfare impacts for either a PGI or PDO. There are findings in the literature on quality assurance schemes and commodity promotion. They are based on either the strong assumption of a market failure in the sense of AKERLOF when computing the consumer benefit (HERRMANN et al., 2002) or they concentrate on producer benefits (ALSTON et al., 2005).

An evidence-based policy analysis on the welfare implications of PGIs and PDOs would have to incorporate the following main aspects:

- Does the GI regulation really reduce quality uncertainty? Several empirical studies suggest that consumers are often not aware of the contents of GI labels and regulation (e. g. FOTOPOULOS and KRYSTALLIS, 2000; TEUBER 2011, pp. 88 et seq.; TREGEAR et al., 1998). Thus, it is very unlikely that a full reduction of quality uncertainty occurs due to GI regulation. Consequently, welfare effects could change strongly. Up to now, this issue has not been incorporated in welfare analyses of GI protection.
- Findings from surveys suggest that consumers prefer foods from their own geographical area as they are willing to support the regional economy (HENSELEIT et al., 2009). This ethnocentrism raises the question whether PGIs or PDOs will shift demand to a level that is justified by a market failure in the AKERLOF sense. Ethnocentrism has not been incorporated in empirical assessment of PGIs and PDOs but might strengthen the position of opponents who argue that GI regulation is a hidden measure of trade protection and can discourage innovation (JOSLING, 2006).
- Is a market failure, i.e. the extreme case of AKERLOF's lemon case, really the appropriate benchmark situation for the evaluation of GI regulation? It appears much more likely that the situation without regulation would be characterized by a market with screening by consumers and signalling by producers which might have some imperfections but would not entirely fail.

Food Taxes as a Health-policy Instrument: Thoughts on Effectiveness and Efficiency

Despite an unprecedented supply of nutritious, safe, and affordable food, many people in industrialized countries follow poor diets that have adverse health effects. The list of lifestyle diseases related to nutrition ranges from obesity to type-II diabetes, psychosocial problems, cardiovascular diseases, and arthrosis to certain types of cancer. Resulting costs for society from increased health care utilization and productivity losses from days of illness are substantial. Therefore, public health experts and health organizations have called for stronger government interventions to approach the problem (e.g. WHO, 2004; BROWNELL and FRIEDEN, 2009; POPKIN, 2009). One of the proposed instruments are fiscal measures, so called "fat taxes", that are levied on energy-dense food products regarded as unhealthy because of their high sugar or fat content. In light of the heated public discussion about such strong measures it is worthwhile to take a scientific perspective on the issue. Thus, we elaborate on the following pages on the objectives of fiscal measures, on theoretical considerations about their effectiveness and efficiency, and on the empirical evidence that has been gathered so far about both justification and expectable impacts.

Figure 4 presents the rationale and expected effects of food taxes as a health-policy instrument. For governments to intervene in markets, economists usually demand the existence of a *market failure*. If there is no market failure, the behaviour of rational and informed individuals acting on free markets should maximize individual utility as well as social welfare (CASH and LACANILAO, 2007; CAWLEY, 2011). One major argument in favour of fat taxes is a negative externality that arises from publicly funded health care. A second rationale results from possible self-control problems that lead to present-biased behaviour preventing people from behaving in a way that maximizes their long-term utility.

Once introduced, the diagram illustrates that the cause-effect chain of fat taxes can be very long. Assumed immediate *effects* on *food markets* comprise changes in the price ratio between unhealthy and healthy foods (BROWNELL, 1994; BATTLE and BROWNELL, 1996) and incentives for producers to reformulate their products in a way that the taxed ingredient is used less. Finally, taxes may generate revenues that could be used to finance information campaigns on eating and health (JACOBSON and BROWNELL, 2000; KUCHLER et al., 2005).

The changed market conditions then provide incentives for consumers to change their *consumption behaviour*. Consumption of the taxed product decreases and people may substitute towards healthier options. People's diets improve and overweight and obesity as well as associated diseases decrease. Healthier people demand less health services, relieving public social systems and individual and *social welfare* rises.

Our evaluation of the evidence base for fat taxes as a health policy in the remainder of this section will proceed as follows. In a first step, we will examine whether the market failures that serve as a rationale for fat taxes are theoretically and empirically justified. We believe that such a discussion is important and necessary, despite the statement of the Scientific Advisory Board that policy goals are the primacy of policymakers. Thereafter, we turn to the empirical evidence on the impacts a fat tax may exert on the subsequent links of the cause-effect chain in Figure 4. Finally, we discuss additional investigations that are necessary for serious evidence-based policy recommendations.

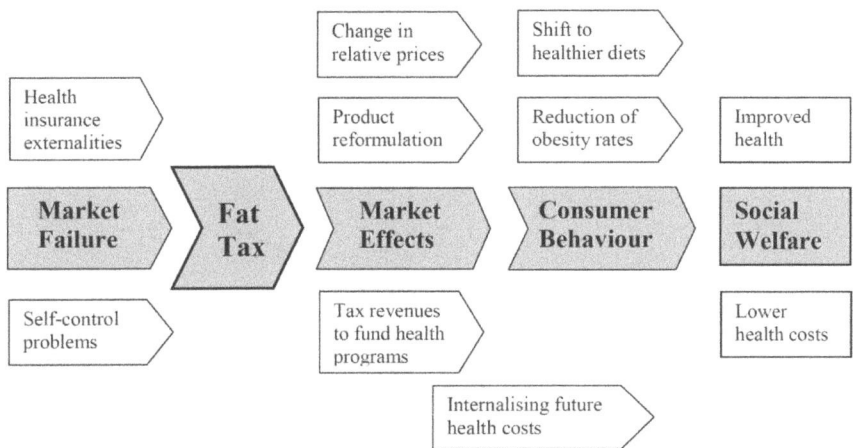

Fig. 4: Rationale and expected impacts of a fat tax
Source: Own presentation

Does empirical evidence support the justification for fiscal measures?
Figure 5 illustrates the market failure argument of negative externalities that emerges from consumption of a certain good, which may be sugar, fat or the amount of energy in our case. The curve *MPB* depicts marginal private benefits or marginal utility from consumption that are a decreasing function of consumption level Q. Additionally, there are marginal private costs of consumption (*MPC*). The positive slope of *MPC* results from possibly increasing medical expenses at higher intakes and higher body weights, although our general argumentation would be the same with constant marginal costs.

Just considering private costs and benefits, the optimal consumption level would be Q_P, at the intersection of *MPB* and *MPC*. If *MPC* and *MPB* comprised all benefits and costs of energy intake and body weight, body weight at consumption Q_P would be socially optimal. This even holds when persons of such a weight would be considered overweight or obese from a medical perspective.

However, *MPC* may not account for all societal costs that arise from intake of energy or sugar. The major example for such costs not borne by consumers directly is public health expenditures to treat illness caused by poor nutrition behaviour. The marginal costs of consumption for the entire society are expressed by *MSC* and are the sum of marginal private costs (*MPC*) and marginal external costs (*MEC*). When *MEC* are positive, socially optimal consumption Q_S would be smaller than privately optimal consumption Q_P resulting in a welfare loss of the triangle ($b + c$). In such a situation, governmental intervention could increase overall welfare by taxing consumption at a rate t that equals *MEC*. External costs would be internalized, resulting in a consumption level of Q_S.

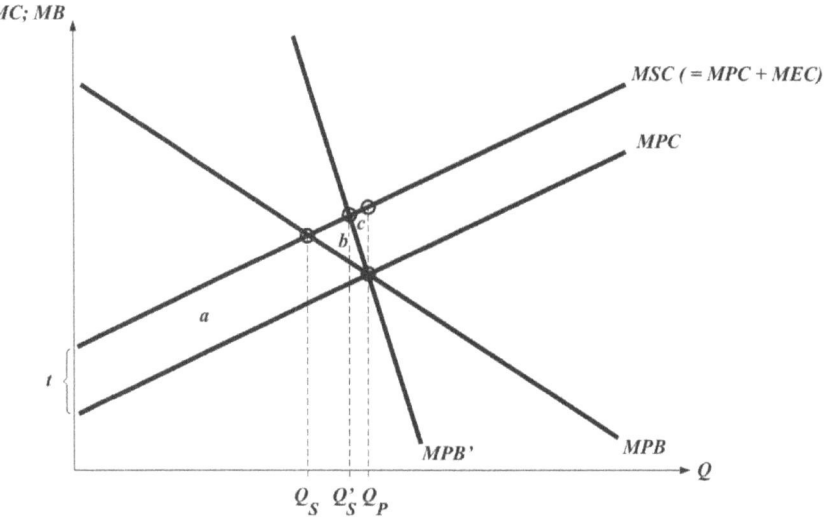

Fig. 5:   Welfare considerations for a consumption externality

Source: Own presentation

Implementing this theoretical solution in practice is hindered by a high degree of uncertainty about the underlying relationships. For example, CASH and LA-CANILAO (2007) emphasize that it is extremely difficult if not impossible to determine the marginal external costs of sugar or fat consumption and, thus, the proper tax rate. Welfare gains would certainly decrease if the calculated tax rate was "wrong". In the extreme case, there may be even negative welfare effects, for instance, in the presence of a uniform tax rate which is too high from the social point of view.

Another substantial argument concerns the measures of economic costs that are commonly used with reference to an externality. These may overstate the welfare loss from bad nutrition behaviour, because they usually refer to the *total* external medical costs described by the area $(a + b + c)$. However, these health care expenditures represent at the same time the revenues of hospitals, physicians or suppliers of pharmaceuticals[2] and are thus not lost to society. Crucial for allocative inefficiency is only the area $(b + c)$. Many authors recurring to the "huge economic costs" of obesity fail to differentiate between a simple redistribution through the health subsidy and allocative inefficiency emerging from moral hazard possibly caused by this subsidy.

The size of the welfare loss strongly depends on the slope of the demand curve (i.e. *MPB*). When demand for food energy or body weight is inelastic as in the case of *MPB'*, the welfare loss shrinks to *c*. BHATTACHARYA and SOOD (2011) argue "that the existing literature [..] suggests that obese people on average do bear the costs and benefits of their eating and exercise habits" (p. 141). Additionally, the authors examine the effects of intensified public health care on obesity on the basis of the RAND Health Insurance Experiment, the "only large-scale randomized health insurance intervention ever conducted" where "each family was assigned a different level of insurance coverage on a random basis" (p. 151). The authors found no significant differences in the level or change of obesity or Body Mass Index (BMI) between people with health plans of different generosity.

Exploiting state-to-state variation in insurance coverage, MARKOWITZ and KELLY (2009) as well as BHATTACHARYA et al. (2011) examine differences along the extensive margin, i.e. between people with and people without insurance, applying instrumental variables approaches. While the former find no effect on obesity, the latter report that "moving from a lack of insurance to Medicaid increases body mass index in the long run by over 2 points" (p. 153). BHATTACHARYA and BUNDORF (2009) find that obese workers with employer-sponsored health insurance have to pay for higher health care costs in the form of lower wages. Hence, in such a private setting, external effects are internalized. BHATTACHARYA and SOOD (2011) conclude that although there are substantial subsidies from thin people to obese people within the health system, this is more a matter of redistribution. The evidence that insurance induces people to behave unhealthily and, thus, for an externality causing inefficiencies is rather weak.

Even in the presence of a substantial welfare loss caused by externalities it would be questionable whether a fat tax is the adequate response. BROWNING (1999) discusses whether external effects mitigated by public institutions like the tax system or public health insurance are *genuine* externalities. ALSTON et al.

---

2 Of course, the figure may exaggerate the area (a + b + c). When MEC are a nonlinear function of Q, external costs emerge only at higher levels of Q.

(2012) argue that "using food policies as obesity policy is an inherently 'second-best' approach because the economic distortion does not stem from a distortion in the price of food" (p. 165). This is in line with CORDEN's (1957) theory of distortions suggesting to take corrective action as close to the problem as possible. In the case of an insurance externality, the first-best solution would be differentiated insurance premiums proportional to a person's risk, what may be "politically infeasible" (ALSTON et al., 2012, p. 165). In such a case, politicians could accept fat taxes as second-best approach or they could even wave any economic welfare considerations and implement taxes in a therapeutic manner. Then they need evidence-based recommendations on the optimal design of a tax with regard to effectiveness and efficiency. We will discuss respective studies below in the following section.

Besides negative externalities, another intensely discussed rationale for taxes is time-inconsistent behaviour which is often labelled as *"self-control problems"*. CUTLER et al. (2003) argue that reduced time costs of food preparation and increased availability affects those individuals the most that possess "hyperbolic discount rates". In this case, short-term preferences are not in accordance with long-term preferences. ETILÉ (2011) concludes that individuals who show hyperbolic discounting "do not receive in the long run the utility they would have received if they were to have been consistent" (p. 729). Some authors state that the presence of hyperbolic discounting is said to justify interventions (BRUNELLO et al., 2009; SASSI, 2010; ETILÉ, 2011). O'DONOGHUE and RABIN discuss "sin taxes" as a means to help people overcome self-control problems (2006). These are intended to make present consumption more expensive and, by this, bring short-term incentives in accordance with long-term objectives. However, ETILÉ also warns to be "cautious, however, in labelling behaviours as 'irrational'" (p. 728) and states that "absent good measures of time preferences in a large-scale food survey, hyperbolic discounting models are not yet testable."

Empirical evidence on the causal effects of taxes on unhealthy foods

Most of the empirical studies analyzing potential effects of fat taxes concentrate on food and nutrient consumption aspects or on impacts on body weight. Only few studies address the issues of a *market reaction* or *social welfare* outcomes depicted in Figure 4. Quite a number of reviews have already provided comprehensive compilations of empirical studies on fat taxes (POWELL and CHALOUPKA, 2009; POWELL et al., 2013; FAULKNER et al., 2011; POWELL and CHRIQUI, 2011; ANDREYEVA et al., 2010; THOW et al., 2010).

One approach frequently used is the estimation of *reduced-form relationships* between prices of selected food products and BMI or the probability of being obese. For the largest part, these studies find significant price effects on body weight on the one hand. On the other hand, however, the received elasticities are usually smaller than 0.1 in absolute values indicating that changing prices for food and beverages have little effect on weight outcomes (POWELL et al., 2013). From a methodological perspective, a main advantage of such price-weight studies is that they provide an easy way to directly assess the impact of exogenous price variation on BMI. Another advantage is the use of fixed-effects (FE) panel models that account for unobserved heterogeneity possibly causing biased coefficients. At the same time, the results of FE models mainly represent short-run effects of prices on BMI, though, and do not offer any information on long-run effects of price changes on body weight. Moreover, these studies neglect that poor nutrition behaviour exerts its health effects not only indirectly via body weight but also directly, when people consume few vitamins or many saturated fatty acids.

In order to get a picture of the expectable consumption patterns after a policy change and their impact on budgets and nutrient patterns that have important implications for welfare and health aspects, ETILÉ (2011) advocates demand-system analyses that contain explicit information on own- and cross-price effects and substitutive relationships. SCHROETER et al. (2008) provide a theoretical assessment of substitutive relationships. Moreover, they show on the basis of existing elasticities from the literature and methods of energy accounting that a tax on food away from home could actually lead to an increase in body weight.

Several *demand-system* studies examine effects of food taxes and subsidies with similar approaches. First, the authors estimate demand systems for one or more food categories. Second, the resulting price and expenditure elasticities are linked with the nutrient content of the included food products to receive nutrient elasticities. Third, simulations are conducted that should reveal how price changes caused by taxes or subsidies alter nutrient intake. In an optional fourth step, simulated changes in energy intake are translated into weight changes.

For France, ALLAIS et al. (2009) find that "a fat-tax policy is unsuitable for substantially affecting the nutrients purchased by French households and leads to ambiguous effects" (p. 243) such as lower intakes of important nutrients like several vitamins and minerals (ALLAIS et al., 2009). They also show that fat taxes would generate substantial revenues while being highly regressive. CHOUINARD et al. (2007) examine a tax on fat content in dairy products and find that a 10%-tax on fat content would reduce fat consumption by less than 1%. This study is one of the few that considers welfare, but only for consumption distortions caused by the tax. A fat-tax would be extremely regressive and elderly and poor people would incur higher welfare losses.

The subject of NORDSTRÖM and THUNSTRÖM's (2011) study is the effect of taxes and subsidies targeted at healthier grain consumption. Results indicate that, regarding tax payments, such reforms are progressive. The authors point out, however, that policies directed at other food groups may have different welfare effects. The authors find that it is the highest income group that increases their fibre consumption the most. At the same time, "the increase in fibre intake is accompanied by substantial increases in the intake of unhealthy nutrients, though, making the net health effects difficult to evaluate" (p. 9). This example mirrors an often pronounced view that policies aimed at altering the consumption of one target nutrient (e.g. saturated fats) or energy may have negative effects on intakes of other nutrients. KUCHLER et al. (2005) estimate single demand equations for snack foods like chips and other salty snacks to assess likely effects of a tax on these food items. Their estimates suggest that taxes would have only minor effects on dietary quality but would generate large revenues.

Since soft drinks are perceived as the single largest contributor to energy intake, many researchers especially in the US identified soft drinks as a promising target for taxes (BROWNELL et al., 2009). Quite high own-price elasticities in literature of around 0.8-1.0 in absolute values promise relatively strong consumer responses to taxes (ANDREYEVA et al., 2010). This notion disregards possible substitution towards other energy-rich beverages, though. FLETCHER et al. (2010) for example, reveal counterintuitive effects of sin taxes on sugar-sweetened beverages. They find that "moderate reductions in soft drink consumption from current soda tax rates" are "completely offset by increases in calories from other beverages" (p. 968), mainly whole milk. Increased consumption of more nutritious substitutes like juices and milk, however, could have positive side effects on health through higher intake of valuable vitamins and minerals (FLETCHER, 2011).

Only few countries have levied special taxes on unhealthy food so far. One exemption is Denmark's tax on saturated fatty acids, which has been evaluated by JENSEN and SMED (2013). They assess the tax effect on fats like butter, butterblends, margarine and oils based on GfK panel data and find that purchases of these fats decreased by 10-15 %. Moreover, they found a churn of customers from high-price supermarkets to low-price discounters. The latter took advantage of this situation by raising prices more than the pure tax increase. The authors warn to interpret their findings with care since the post-tax period examined is very short (10 months). Despite the implementation of the tax in real life, the Danish example cannot be regarded as classic experiment since the policy affects all Danes and we have no control group for that treatment. Another critical point in their analysis is the choice of products considered. Products in the category fats may have the highest content of fatty acids, however, meat and milk products as well as processed foods surely account for the largest part of daily fat intake. Thus, results are not representative for overall intake.

The only study that addresses *cost-effectiveness* of fat taxes has been conducted by SASSI et al. (2009) from the OECD. They find that "fiscal measures generate reductions in health care expenditure which more than offset intervention costs, thus leading to savings of about [...] $ 32.6 billion" (ibid., p. 43). However, these results are based on simulation models with numerous assumptions on parameters and costs and may vary considerably in practice.

A weak evidence base for fat taxes as health-policy instruments

Theoretical considerations and empirical evidence presented above strongly suggest that taxes on foods are neither an effective nor an efficient health-policy instrument. Many issues and relationships along the cause-effect chain in Figure 4 remain uncertain and there are considerable doubts that even the most sophisticated future research methods can ever provide an exact assessment of the underlying complex relationships. The following aspects raise questions in particular:

- We have to be aware of the LUCAS critique regarding the measured consumer reactions to prices since we do not know how these will change when the policy is introduced (LUCAS, 1976). Taxed foods could be increasingly stigmatized in society resulting in stronger avoidance. In contrast, consumers may also turn to protest behaviour and buy the taxed foods in a "now-more-than-ever" manner.
- Evidence on responses of supply-side actors to fat taxes is completely lacking. Large parts of the simulations or evidence on price reactions discussed above are pointless, if we had no knowledge on how market prices change in response to a tax. Industrial organization studies generally find a high degree of price rigidity in retailing (HERRMANN et al., 2005). Moreover, it is uncertain how producers will react in terms of product reformulation, whether products will get healthier or whether firms provide products of lower quality to compensate for higher prices.
- Theory and evidence on how fat taxes affect welfare and utility of consumers exactly are insufficient. Trade-offs between welfare losses through consumption restrictions and welfare gains through (an assumed) weight loss are inadequately described by static models based on neoclassics and lack statements on dynamic adjustment processes.

Discussion and Conclusions

Proposals towards evidence-based policy analyses and recommendations are a challenge for agricultural and food policy. They are important as policy advice based on empirical impact analysis is crucial for the researchers' credibility.

It is questionable, however, whether evidence-based policy analysis and advice should generally rely on the experimental approach to program evaluation (WISSENSCHAFTLICHER BEIRAT, 2013). There is no doubt that the experimental approach does extend the toolbox of agricultural and food policy analysis significantly. Applications of the experimental or quasi-experimental approach have been fruitfully applied to the analyses of consumer reactions to safer food (LUSK and SHOGREN, 2007) and to regional impacts of Common Agricultural Policy (PETRICK and ZIER, 2011). There are further agricultural and food policies which affect the whole market, i.e. all producers and consumers, and make it much more difficult to distinguish control and treatment groups as in medical sciences. In all these cases, estimated structural models can be used in agricultural policy analysis as in the past. Thus, the experimental approach to program evaluation will not always be the first-best option of empirical policy analysis. There is no need to forfeit the application of structural models based on economic theory that have dominated quantitative analyses in agricultural economics over decades. A best-practice approach with regard to the choice of methodology seems superior or, "building bridges between structural and program evaluation approaches", as HECKMAN (2010) puts it.

New food policy instruments which are targeted at food quality and health are still far away from evidence-based analysis and recommendations. We showed this for the protection of geographically differentiated foods and for food taxes. The protection of geographical indications for agricultural foodstuffs with a strong link between geographical origin and product quality is often motivated by AKERLOF's lemon case. But it is in most cases not investigated whether a market failure with information asymmetry and quality uncertainty by consumers would actually be given in the hypothetical case without regulation. The potential of screening and signaling as market-conform options for consumers or producers to distinguish high- and low-quality products is often not considered any more. For PDOs and PGIs, price incentives for producers are sometimes inferred directly from price premia consumers are paying compared to a lower-quality product of the mass market. Only very rarely, the producer price incentives are computed on a theory-based framework. More evidence-based analysis is needed here. Empirical studies on the income and welfare effects of PDOs and PGIs are also lacking including analyses of (i) the costs of participation in the producer club under GI regulation and (ii) evidence on whether and how regulation actually affects quality uncertainty and search costs by consumers.

Substantial doubts remain also for food taxes as a health-policy instrument. First, the case of market failure by negative externalities as a rationale for taxes is much less clear than commonly spread. The same is true for the case of self-control problems which we have not discussed in more detail here. Second, empirical evidence on the effects of taxes shows very modest effects on intake of energy or BMI. This may be due to various substitution possibilities that exist in highly differentiated food markets where consumption decisions depend to a large degree on non-pecuniary factors. A lesson for future (economic) research is that increasing efforts should be directed to the underlying reasons of a problem first, before conducting massive research on effects of a policy that does not address the genuine source of the problem.

List of Referneces

Akerlof, G. (1970): The Market for 'Lemons': Quality Uncertainty and the Market Mechanism. Quarterly Journal of Economics, 84 (3), 488-500.

Allais, O., Bertail, P. and Nichèle, V. (2009): The Effects of a Fat Tax on French Households' Purchases: A Nutritional Approach. American Journal of Agricultural Economics, 92 (1), 228-245.

Alston, J.M., Chalfant, J.A, Christian, J.E., Meng, E. and Piggott, N. (2005): The Benefits and Costs of Promotion Programs for California Table Grapes. In: Kaiser, H.M., J.M. Alston, J.M. Crespi and R.J. Sexton (eds.). The Economics of Commodity Promotion Programs. Lessons from California. New York: Peter Lang, 71-94.

Alston, J.M., Okrent, A. and Parks, J.C. (2012): U.S. Food Policy and Obesity. In: Maddock, J. (ed.), Public Health - Social and Behavioral Health, InTech, 165-184.

Alston, J.M. and James, J.S. (2002): The Incidence of Agricultural Policy. In: Gardner, B.L. and G.C. Rausser (eds.), Handbook of Agricultural Economics, 2B, Agricultural and Food Policy, Amsterdam Elsevier, 1690-1749.

Alston, J.M., Norton, G.W. and Pardey, P.G. (1995): Science Under Scarcity – Principles and Practice for Agricultural Research Evaluation and Priority Setting. Ithaca, NY, Cornell University Press.

Andreyeva, T., Long, M. and Brownell, K.D. (2010): The Impact of Food Prices on Consumption: A Systematic Review of Research on Price Elasticity of Demand for Food. American Journal of Public Health, 100 (2), 216–222.

Anderson, K. (2010): Krueger, Schiff and Valdés Revisited: Agricultural Price and Trade Policy Reform since 1960. Applied Economic Perspectives and Policy, 32 (2), 195-231.

Angrist, J.D. and Pischke, J.-S. (2010): The Credibility Revolution in Empirical Economics: How Better Research Design is Taking the Con out of Econometrics. Journal of Economic Perspectives, 24 (2), 3-30.

Antle, J. (2001): Economic Analysis of Food Safety. In: Gardner, B.L. and G.C. Rausser (eds.), Handbook of Agricultural Economics, 1B, Marketing, Distribution and Consumers. Amsterdam, Elsevier, 1083-1136.
Battle, E.K. and Brownell, K.D. (1996): Confronting a rising tide of eating disorders and obesity: Treatment vs. prevention and policy. Addictive Behaviors, 21 (6), 755-765.
Becker, T. (2009): European Food Quality Policy: The Importance of Geographical Indications, Organic Certification and Food Quality Assurance Schemes in European Countries. Estey Center Journal of International Law and Trade Policy, 10 (1), 111-130.
Bessler, D.A., Dorfman, J.H, Holt, M.T. and Lafrance, J.T. (2010): Econometric Developments in Agricultural and Resource Economics: The First 100 Years. American Journal of Agricultural Economics, 92 (2), 571-589.
Bhattacharya, J. and Bundorf, M.K. (2009): The Incidence of the Healthcare Costs of Obesity. Journal of Health Economics, 28 (3), 649-58.
Bhattacharya, J., Bundorf, K., Pace, N. and Sood, N. (2011): Does Health Insurance Make You Fat? In: Grossman, M. and N. H. Mocan (eds.), Economic Aspects of Obesity, NBER Books, Chicago, University of Chicago Press, 35-64.
Bhattacharya, J. and Sood, N. (2011): Who Pays for Obesity? Journal of Economic Perspectives, 25 (1), 139-158.
Bigman, D. (1982): Coping with Hunger: Toward a System of Food Security and Price Stabilization. Cambridge, Mass., Ballinger.
Brownell, K.D. (1994): Get Slim with Higher Taxes. The New York Times, December 15, 1994.
Brownell, K.D., Farley, T., Willet, W.C., Popkin, B.M., Chaloupka, F.J., Thompson, J.W. and Ludwig, D.S. (2009): The Public Health and Economic Benefits of Taxing Sugar-Sweetened Beverages. The New England Journal of Medicine, 361 (16), 1599-1605.
Brownell, K.D. and Frieden, T.R. (2009): Ounces of Prevention - The Public Policy Case for Taxes on Sugared Beverages. The New England Journal of Medicine, 360 (18), 1805-1808.
Browning, E.K. (1999): The Myth of Fiscal Externalities. Public Finance Review, 27 (1), 3-18.
Brunello, G., Michaud, P.-C. and Sanz-de-Galdeano, A. (2009): The Rise of Obesity in Europe – an Economic Perspective. Economic Policy, 24 (59), 551-596.
Cash, S.B. and Lacanilao, R.D. (2007): Taxing Food to Improve Health: Economic Evidence and Arguments. Agricultural and Resource Economics Review, 36 (2), 174-182.
Cawley, J. (2011): The Economics of Obesity. In: Cawley, J. (ed.), The Oxford Handbook of the Social Science of Obesity, New York, Oxford University Press, 120-137.
Chouinard, H.H., Davis, D.E., LaFrance, J.T. and Perloff, J.M. (2007): Fat Taxes: Big Money for Small Change. Forum for Health Economics & Policy, 10 (2).
Corden, W.M. (1957): Tariffs, Subsidies and the Terms of Trade. Economica, 24, 235-242.
Cutler, D.M., Glaeser, E.L. and Shapiro, J.M. (2003): Why Have Americans Become More Obese? Journal of Economic Perspectives, 17 (3), 93-118.
Deselnicu, O.C., Costanigro, M., Souza-Monteiro, D.M. and McFadden D.T. (2013): A Meta-Analysis of Geographical Indication of Food Valuation Studies: What Drives the Premium for Origin-Based Labels? Jornal of Agricultural and Resource Economics, 38 (2), 204-219.

Etilé, F. (2011): Food Consumption and Health. In: Lusk, J.L., J. Roosen and J.F. Shogren (eds.), The Economics of Food Consumption and Policy, New York, Oxford University Press, 716-746.
European Commission (2008): Evaluation of the CAP Policy on Protected Designations of Origin (PDO) and Protected Geographical Indications (PGI). Final Report, London Economics. Available online at: http://ec.europa.eu/agriculture/eval/reports/pdopgi/report_en.pdf (accessed March 11, 2014).
European Commission (2012): Regulation (EU) No 1151/2012 of the European Parliament and of the Council of 21 November 2012 on Quality Schemes for Agricultural Products and Foodstuffs. Official Journal of the European Union, L343/1-L343/29.
Faulkner, G.E.J., Grootendorst, P., Nguyen, V.H., Andreyeva, T., Arbour-Nicitopoulos, K., Auld, M.C., Cash, S.B., Cawley, J., Donnelly, P., Drewnowski, A., Dubé, L., Ferrence, R., Janssen, I., LaFrance, J., Lakdawalla, D., Mendelsen, R., Powell, L.M., Traill, W.B. and Windmeijer, F. (2011): Economic Instruments for Obesity Prevention: Results of a Scoping Review and Modified Delphi Survey. International Journal of Behavioural Nutrition and Physical Activity, 8 (109).
Fletcher, J.M. (2011): Soda Taxes and Substitution Effects: Will Obesity Be Affected? Choices, 26.
Fletcher, J.M., Frisvold, D.E. and Tefft, N. (2010): The Effects of Soft Drink Taxes on Child and Adolescent Consumption and Weight Outcomes. Journal of Public Economics, 94 (12), 967-974.
Fotopoulos, C. and Krystallis, A. (2000): Quality Labels as a Marketing Advantage: The Case of the 'PDO Zagora' Apples in the Greek Market. European Journal of Marketing, 37 (10), 1350-1374.
Fox, K. (1986): Agricultural Economists as World Leaders in Applied Econometrics, 1917 33. American Journal of Agricultural Economists, 68 (2), 381-386.
Gardner, B.L. (1987): The Economics of Agricultural Policies. New York, NY, MacMillan.
Giersch, H. (1990): On Being a Public Economist. Lecture at the Price-awarding Ceremony of the Paolo Baffi International Price of Economics 1989, Kiel.
Heckman, J.J. (2010): Building Bridges between Structural and Program Evaluation Approaches to Evaluating Policy. Journal of Economic Literature, XLVIII (2), 356-398.
Henseleit, M., Kubitzki, S. and Teuber, R. (2009): Determinants of Consumer Preferences for Regional Food Products. In: Canavari, M., N. Cantore, A. Castellini, E. Pignatti and R. Spadoni (eds.), International Marketing and Trade of Quality Food Products. Wageningen, Wageningen Academic Publishers, 263-278.
Herrmann, R., Möser, A. and Weber, S. (2005): Price Rigidity in the German Grocery-Retailing Sector: Scanner-Data Evidence on Magnitude and Causes. Journal of Agricultural & Food Industrial Organization, 3 (1), Article 4.
Herrmann, R. and Schmitz, P.M. (1984): Stabilizing Producers' Revenue by Fixing Agricultural Prices within the EC? European Review of Agricultural Economics, 11 (4), 395-414.
Herrmann, R. and Teuber, R. (2011): Geographically Differentiated Products. In: Lusk, J.L., J. Roosen and J.F. Shogren (eds.), The Oxford Handbook of the Economics of Food Consumption and Policy. Oxford, Oxford University Press, 811-842.

Herrmann, R., Thompson, S. and Krischik-Bautz, S. (2002): Bovine Spongiform Encephalopathy and Generic Promotion of Beef: An Analysis for 'Quality from Bavaria'. Agribusiness, 18 (3), 369-385.

Houck, J.P. (1986): Elements of Agricultural Trade Policies. New York, NJ, MacMillan Publishing Company.

Imbens, G.W. and Wooldridge, J.M. (2009): Recent Developments in the Econometrics of Program Evaluation. Journal of Economic Literature, 47 (1), 5-86.

Ittersum, van K., Meulenberg, M.T.G., Trijp, van H.C.M. and Candel, M.J.J.M. (2007): Consumers' Appreciation of Regional Certification Labels - A Pan-European Study. Journal of Agricultural Economics, 58 (1), 1-23.

Jacobson, M.F. and Brownell, K.D. (2000): Small Taxes on Soft Drinks and Snack Foods to Promote Health. American Journal of Public Health, 90 (6), 854-857.

Jensen, P.H. (2013): What is Evidence-Based Policy? Melbourne Institute Policy Briefs Series, Policy Brief No. 4/13, Faculty of Business & Economics, The University of Melbourne.

Jensen, J.D. and Smed, S. (2013): The Danish Tax on Saturated Fat – Short Run Effects on Consumption, Substitution Patterns and Consumer Prices of Fats. Food Policy, 42, 18-31.

Josling, T. (2006): The War on Terroir: Geographical Indications as a Transatlantic Trade Conflict. Journal of Agricultural Economics, 57 (3), 337-363.

Just, R.E., Hueth, D.L. and Schmitz, A. (2004): The Welfare Economics of Public Policy: A Practical Approach to Project and Policy Evaluation. Northampton, Mass., Edward Elgar.

Kaiser, H.M., Alston, J.M., Crespi, J.M. and Sexton, R.J. (2005): The Economics of Commodity Promotion Programs. Lessons from California. New York, Peter Lang.

Koester, U. and Schmitz, P.M. (1982): The EC Sugar Market Policy and Developing Countries. European Review of Agricultural Economics, 9 (2), 183-204.

Kuchler, F., Tegene, A. and Harris, J.M. (2005): Taxing Snack Foods: Manipulating Diet Quality or Financing Information Programs? Review of Agricultural Economics, 27 (1), 1-17.

Lee, Y.-J. (2005): Micro-Econometrics for Policy, Program and Treatment Effects. (Advanced Texts in Econometrics), Oxford, Oxford University Press.

Lence, S.H., Marette, S., Dermot, J.H. and Foster, W. (2007): Collective Marketing Arrangements for Geographically Differentiated Agricultural Products – Welfare Impacts and Policy Implications. American Journal of Agricultural Economics, 89 (4), 947-963.

Lloyd, P., Croser, J.L. and Anderson, K. (2010): Global Distortions to Agricultural Markets – Indicators of Trade and Welfare Impacts, 1960 to 2007. Review of Development Economics, 14 (2), 141-160.

Lucas, R.E. (1976): Econometric Policy Evaluation: A Critique. Carnegie-Rochester Conference Series on Public Policy, 1 (1), 19-46.

Lusk, J.L. and Shogren, J.F. (2007): Experimental Auctions: Methods and Applications in Economic and marketing Research. Cambridge, Cambridge University Press.

Markowitz, S. and Rashad Kelly, I. (2009): Incentives in Obesity and Health Insurance. Inquiry, 46 (4), 418-32.

Moschini, G.C., Menapace, L. and Pick, D. (2008): Geographical Indications and the Competitive Provision of Quality in Agricultural Markets. American Journal of agricultural Economics, 90 (3), 794-812.

Myers, R.J., Sexton, R.J. and Tomek, W.G. (2010): A Century of Research on Agricultural Markets. American Journal of Agricultural Economics, 92 (2), 376-402.

Newbery, D.M.G. and Stiglitz, J.E. (1981): The Theory of Commodity Price Stabilization – A Study in the Economics of Risk. Oxford, Clarendon Press.
Nordström, J. and Thunström, L. (2011): Can Targeted Food Taxes and Subsidies Improve the Diet? Distributional Effects Among Income Groups. Food Policy, 36 (2), 259-271.
O'Donoghue, T. and Rabin, M. (2006): Optimal Sin Taxes. Journal of Public Economics, 90 (10), 1825-1849.
Petrick, M. and Zier, P. (2011): Regional Employment Impacts of the Common Agricultural Policy Measures in Eastern Germany: A Difference-in-Difference Approach. Agricultural Economics, 42 (2), 183-193.
Popkin, B.M. (2009): The World is Fat. The Fads, Trends and Policies, that are Fattening the Human Race. New York, Avery.
Powell, L.M. and Chaloupka, F.J. (2009): Food Prices and Obesity – Evidence and Policy Implications for Taxes and Subsidies. The Milbank Quarterly, 87 (1), 229-257.
Powell, L.M. and Chriqui, J.F. (2011): Food Taxes and Subsidies: Evidence and Policies. In: Cawley, J. (ed.), The Oxford Handbook of the Social Science of Obesity, New York, Oxford University Press, 639-664.
Powell, L.M., Chriqui, J.F., Khan, T., Wada, R. and Chaloupka, F.J. (2013): Assessing the Potential Effectiveness of Food and Beverage Taxes and Subsidies for Improving Public Health – A Systematic Review of Prices, Demand and Body Weight Outcomes. Obesity Reviews, 14 (2), 110-128.
Rausser, G.C., Swinnen, J.F.M. and Zusman, P. (2011): Political Power and Economic Policy. Theory, Analysis and Empirical Applications. Cambridge, Mass., Cambridge University Press.
Rosen, S. (1974): Hedonic Prices and Implicit Markets: Product Differentiation in Pure Competition. Journal of Political Economy, 82 (1), 34-55.
Sackett, D.L., Rosenberg, W.M.C., Gray, J.A.M., Haynes, R.B. and Richardson W.S. (1996): Evidence Based Medicine – What It Is and What It Isn't. British Medical Journal, 312, 13 January, 71-72.
Sassi, F., Cecchini, M., Lauer, J. and Chisholm D. (2009): Improving Lifestyles, Tackling Obesity – The Health and Economic Impact of Prevention Strategies. OECD Health Working Papers, No. 48, Paris, OECD Publishing.
Sassi, F. (2010): Obesity and the Economics of Prevention. Fit not Fat. Paris: OECD Publishing.
Schiff, M. and Valdés, A. (1992): The Political Economy of Agricultural Pricing Policy. Vol. 4, A Synthesis of the Economics in Developing Countries. (A World Bank Comparative Study), Baltimore, London, The Johns Hopkins Press.
Schmitz, P.M. (1984): Handelsbeschränkungen und Instabilität auf Weltagrarmärkten. (Weltwirtschaftliche Studien 21), Göttingen, Vandenhoeck & Ruprecht.
Schmitz, P.M. and Hesse, J.W. (2009): Das verfassungsrechtliche Aus des Absatzfonds - ökonomische Bewertung und Entwurf einer Nachfolgelösung. Agribusiness-Forschung 23, Institut für Agribusiness, Giessen.
Schroeter, C., Lusk, J. and Tyner, W. (2008): Determining the Impact of Food Price and Income Changes on Body Weight. Journal of Health Economics, 27 (1), 45-68.
Sumner, D.A., Alston, J.M. and Glauber, J.W. (2010): Evolution of the Economics of Agricultural Policy. American Journal of Agricultural Economics, 92 (2), 403-423.

Teuber, R. (2011): The Economics of Geographically Differentiated Agri-Food Products. Theoretical Considerations and Empirical Evidence. Ph.D. Thesis, Justus Liebig University Giessen. Available online at: http://geb.uni-giessen.de/geb/volltexte/2011/8105/pdf/TeuberRamona_2011_03_18.pdf (accessed March 17, 2014).

Tregear, A., Kuznesof, S. and Moxey, A. (1998): Policy Initiatives for Regional Foods: Some insights from Consumer Research. Food Policy, 23 (5), 383-394.

Tyers, R. and K. Anderson (1992), Disarray in World Food Markets: A Quantitative Assessment. Cambridge, Cambridge University Press.

Thow, A.M., Jan, S., Leeder, S. and Swinburn, B. (2010): The Effect of Fiscal Policy on Diet, Obesity and Chronic Disease: A Systematic Review. Bulletin of the World Health Organization, 88 (8), 609-614.

WHO (2004): Global Strategy on Diet, Physical Activity and Health. Available online at: http://www.who.int/dietphysicalactivity/strategy/eb11344/en/index.html (accessed June 1, 2013).

Will, S. (2013): Evaluierung eines Qualitäts- und Herkunftszeichens: Das Beispiel „Geprüfte Qualität - HESSEN". Ph.D. Thesis, Justus Liebig University Giessen. Available online at: http://geb.uni-giessen.de/geb/volltexte/2013/10422/pdf/WillSabine_2013_07_03.pdf (accessed March 17, 2014).

Wissenschaftlicher Beirat (2011): Wissenschaftlicher Beirat beim Bundesministerium für Ernährung, Landwirtschaft und Verbraucherschutz, Risiko- und Krisenmanagement in der Landwirtschaft. Zur Rolle des Staates beim Umgang mit Ertrags- und Preisrisiken. Berichte über Landwirtschaft, 89 (2), 177-203.

Wissenschaftlicher Beirat (2013): Evaluierung wirtschaftspolitischer Fördermaßnahmen als Element einer evidenzbasierten Wirtschaftspolitik. Gutachten des Wissenschaftlichen Beirats beim Bundesministerium für Wirtschaft und Energie, Berlin.

Wissenschaftlicher Beirat Verbraucher- und Ernährungspolitik (2013), Evidenzbasierung ermöglichen! Auf dem Weg zu einer realitätsnahen und empirisch fundierten Verbraucherpolitik. Stellungnahme des wissenschaftlichen Beirats Verbraucher- und Ernährungspolitik beim BMELV, Berlin.

Zago, A.M. and Pick, D. (2004): Labeling Policies in Food Markets: Private Incentives, Public Interventions, and Welfare Effects. Journal of Agricultural and Resource Economics, 29 (1), 150-165.

Regulation of Agricultural Derivatives Markets

Michaela KUHL

Abstract

In view of the instability due to the international financial crisis, European politics has set itself the goal to increase transparency and to strengthen regulation in derivatives markets. This also applies to transactions in agricultural derivatives, despite numerous international studies having found no consensus on the effect of such transactions on the level and volatility of agricultural prices. By the autumn of 2013, many legal steps have already been taken towards a stricter regulation. Many regulations are however still in the process of legislation. The implementation of these measures will take even longer.

Introduction

Since the beginning of the financial crisis, which began with the collapse of Lehman Brothers, debate has flourished on efficient regulation of the financial markets, and how a similar crisis can be avoided in the future. Until the first half of 2008, the prices of many agricultural commodities, as well as oil prices, exploded almost simultaneously, and in several countries led to massive protests of consumers suffering from more expensive staple foods. On the agricultural markets more attention was drawn to the impact of the sharp increase in market participation of financial investors, the so-called "financialization" of commodity markets. In the public, voices were quickly heard, who demanded, like a number of nongovernmental organizations, a limitation or prohibition of speculative activities on the exchanges. From the beginning, there were also individuals and institutions warning against over-hasty conclusions and throwing the baby out with the bath water. Not least in Europe the need for the proper functioning of commodity derivatives markets is seen in light of the progressive loosening of the agricultural political control on agricultural markets. A derivative means a financial instrument, whose price is based on the development of the prices of one or more products ("underlying"). A typical example is a standardized futures contract, in which a specific amount of wheat is sold or bought at a specific date at an agreed price. Besides futures other examples of derivatives are forwards, options and swaps, which may relate to diverse "underlyings" such as various commodities, currencies, stocks or bonds.

Among the agricultural derivatives markets, the commodity futures exchanges play a crucial role. On the one hand, their task is pricing, providing a basis for spot-market transactions. On the other hand, they offer hedging opportunities against price fluctuations and function as an insurance for producers and users of agricultural goods („hedging"). At the same time financial investors are also active on these markets. Usually, they are divided into index investors (index traders) and money managers. Index investors, such as pension funds, hold a long-term commitment (passive strategy) and mostly count on future price increases. However, money managers, such as hedge funds, pursue a more active and short-term oriented strategy; the orientation on rising, but also on falling prices may vary. Money managers are typically described as „speculators". But there are also gray zones, where big physical traders hold speculative positions in addition to hedging, and vice versa, financial actors become shareholders of trading companies and as such engage in physical trade of agricultural commodities.

Studies on the effect of financial transactions on agricultural prices

Meanwhile, a number of studies have been published examining the relationship between the profit-oriented financial transactions with no interest in the underlying physical product and price formation on the market for these products. WILL et.al. (2012) in their comprehensive literature survey of studies published between 2010 and 2012 conclude that most of the studies find no significant effect of index investors' engagements on the level of agricultural prices. Four articles that were released in prestigious Journals after a review process address the issue of price volatility. They find mixed results on the effect of financial speculation on *price volatility*. However, none of the studies found a significant long-term volatility increase caused by the presence of index traders. Nevertheless, the Journal articles are strongly dominated by a few authors; particularly the names of SCOTT, IRWIN and DWIGHT. SANDLER appear frequently. Also the vast number of studies from the gray literature found no significant effect of speculation on price volatility. The conclusions as to whether speculation has significant effect on price level, are split equally, but many methodological limitations are observed in the studies with a critical view on speculation. According to WILL et.al. (2012), the vast majority of the papers therefore "do not confirm that financial speculation increases the price level of agricultural commodities" (p. 19), and most empirical studies also "do not agree that financial speculation increases the price volatility of agricultural commodities" (p. 15). Meanwhile, a series of new studies has been published: ETIENNE, IRWIN and GARCIA (2013) find in their study that bubbles are found frequently in the price patterns of many agricultural products in the years between 1971 and 2011, especially during the periods 1971-76 and 2006-11. According to their analysis, these bubbles were more frequent and longer-lasting in the first period than in the second one. Since in the first period the role of financial actors was still insignificant, they conclude that they cannot be made responsible for the bubbles in the later period either. In an empirical analysis of price developments on corn and soybean markets, using eight different variables, SCHMITZ and MOLEVA (2013) also find no evidence for an influence of speculation on the corn market. On the soybean market they only find a negligible effect (p. 88). In contrast, GUILLEMINOT, OHANA and OHANA (2013) see a significant impact of the financial actors, especially in times of market turbulence as they find increased synchrony in the behaviour of index actors and speculative traders in these times. As an interim conclusion it should be noted that the majority of studies see indeed no reason to hold financial actors responsible for excessive price movements in agricultural markets; however some studies disagree with this conclusion. In this uncertain environment, many politicians are especially interested in how influential international organizations assess the impact of financial investments. In this respect, the EU might find its regulation efforts confirmed by the World Bank: "While it is unlikely that these investments affect long-term price trends, they have most likely affected price variability" (WORLD BANK, 2013, p. 18).

Objectives and state of the EU regulatory proposals as of autumn 2013

Given inconsistent research results and significant public pressure on the political actors - not least by non-governmental organizations such as Foodwatch[1] and Oxfam - several measures were encouraged and initiated at regional and international level over recent years in order to reduce any possible risks of speculative activities on commodity exchanges. For reputational reasons many banks and funds have withdrawn from the market for speculative agricultural derivatives. This decision was certainly made easier due to the declining profitability of this segment.

The political start for stricter regulation of financial markets (not only with regard to agricultural commodities) was the meeting of the G20 in Washington in November 2008, where three main objectives of reform were addressed: improving market transparency, integration of so called over-the-counter (OTC) trade into the regulation and protection against market abuse. The outline of regulation of OTC markets were explicitly formulated in the declaration of the G20 in Pitts-

---

1 In November 2013 a further study initiated by Foodwatch (BASS, 2013) was published. It draws different conclusions from the existing literature than WILL et.al (2012). The author sees an urgent need for "regulatory measures against the excessive speculation in food".

burgh in September 2009. From the beginning regulation efforts were complicated by the fact that a predominant and ever-increasing part of the derivative transactions does not occur on formal exchanges, but are so called over-the-counter (OTC) transactions. Approximately 95% of the derivative transactions do not take place on official exchanges, but are concluded bilaterally between two parties (BIS, 2013). The massive growth of the OTC derivatives market is not surprising, since the products traded are more flexible and can be adapted individually to the user's needs. In contrast, products on the stock exchange are highly standardized, such as futures contracts with a determined size and maturity. However, it is an undisputed advantage of exchanges that they provide a higher level of transparency and reduce the risk of information asymmetries.

The massive increase in derivative transactions in the years before the crisis has drawn the attention to this less transparent part of the financial system. This also applies to the commodity markets, although commodity derivatives represent only a small share of total derivative transactions and seem to be almost insignificant in comparison to interest rate and currency derivatives[2]. Critics argue that the lack of transparency in OTC transactions together with a potentially weak risk management on the part of trading companies with respect to counterparty default risk as well as strong bilateral links could carry a risk of contagion and thus be a source of systemic risk.

Again and again political leaders emphasize that the aim of regulation is to prevent excessive price increases and volatility by enhancing market transparency and efficiency[3]. To this end various measures are discussed, the most important for agricultural futures markets being stronger regulation of OTC markets, reporting requirements and position limits. Financial market regulation in Europe relevant for agricultural commodities is found in several EU directives.

---

2 An analysis of more than 200 industrial enterprises in Germany conducted by the Deutsches Aktieninstitut showed: Industrial companies mainly use taylor-made OTC derivatives, in order to do manage the risks of underlying transactions in the best possible way. Simply structured financial derivatives are preferred. About nine out of ten companies declare that forward transactions such as currency forwards or interest rate swaps represent about 50 percent of their total derivative volume. Compared to other business risks, hedging is not very common in commodities transactions.

3 The German Federal Government puts it this way: „...Strict regulations and transparency requirements will prevent destabilising effects on food prices without impeding the ability of commodity derivatives markets to hedge against risks"(Federal Ministry of Finance, 2013).

European Market Infrastructure Regulation (EMIR) - EU Regulation on OTC derivatives, central counterparties and trade repositories

In summer 2012, the so-called EMIR (European Market Infrastructure Regulation) entered into force. This implemented the objective of the G-20 for increased transparency and regulation of financial markets. In future, standardized OTC transactions will be subject to central clearing and documented in trade repositories. The clearing obligation aims at eliminating the counterparty default risk for the individual market participants and ensures that collateral is provided in a consistent and timely manner. The clearing house operates as an intermediate buyer and seller in each derivatives transaction. The obligation to report every transaction to a trade repository aims at increasing market transparency, with which the authorities hope to quickly detect any financial turmoil and its sources.

Which kind of transactions must meet the clearing obligation and be reported to trade repositories is determined by the ESMA (European Securities and Markets Authority). Its task is the creation of a single and uniform European financial market. If the reporting requirement holds for a certain kind of trade, the counterparties involved, the nature and scope of transaction, the date of maturity, the settlement date of the contracts and the nominal value should be included.

For central clearing to be feasible, a minimum level of standardization is necessary. Implementation is of key importance. In addition, the line between standardized and non-standardized transactions will not be easy to draw. At the same time, for those transactions which are less standardized and therefore not suited for central clearing, higher safety standards will be imposed (such as higher "margin requirements"). Companies must consider whether given a specific problem (i) an expensive tailor-made, (ii) a cheaper but less individual, more standardized solution or (iii) no derivatives transaction at all is most appropriate in their case. A significant share of interest-rate derivative transactions could already be settled through central clearing – but currently few transactions are settled in this way - ; for commodity derivatives central clearing can be regarded as negligible at present: "Commodity (...) derivatives clearing meanwhile is yet to develop..." (KAYA, 2013, p. 16).

The new requirements for risk and collateral management (in particular management of operational and counterparty default risks, appropriate exchange of collateral, etc.) as well as the electronic reporting requirement to trade repositories apply to *all* businesses with OTC transactions. However, the obligation for central clearing is not always binding.

For companies that are not financial institutions ("non-financial corporation", NFC), there are exemptions to the clearing obligation. This aims at avoiding rising hedging costs for the real underlying business of such companies. Below a certain clearing threshold NFCs do not have to clear centrally. In case of commodity derivatives, a threshold of three billion Euros calculated as a moving average over 30 days was set by ESMA. If a NFC exceeds this threshold, then it is treated as a financial institution and has to clear centrally. It is then referred to as NFC+. Those OTC derivative contracts that fall under the hedging definition, i.e., that reduce the risks of real underlying commodity transactions, are not included in the calculation. Currently, no non-financial company in Germany exceeds the clearing threshold, i.e., there are no so called NFC+. However, many larger companies with more than 100 million Euros of total hedging volume or more than 100 OTC derivatives transactions have to bite the bullet and go through an annual external audit process to prove that their volume of derivatives transactions is actually lower than the clearing threshold.

The EMIR regulation is directly applicable in Germany without the need for a national law. But still, laws and regulations need to be adapted at the national level. In Germany, a corresponding EMIR Implementation Act entered into force in February 2013. The competent Authority in Germany is the German Federal Financial Supervisory Authority (BaFin). Currently, clearing houses and trade repositories are being introduced gradually. They must be approved by the BaFin. After having been delayed by several weeks, the start date for reporting derivative transactions is now February 12, 2014. According to ESMA the final details of the clearing requirements should be presented to the EU Commission no later than 15 September 2014.

Markets in Financial Instruments Directive (MiFID II) – An update on the 2007 MiFID Regulation

The MiFID directive of 2007 created a series of deregulation measures aimed towards creating a single European financial market in order to increase competition in financial services. MiFID II relates to the trade activities on orderly trading platforms, such as exchanges, and applies to all financial instruments that do not fall under EMIR regulation. In this context, next to documentation and publication requirements the discussion about position limits also comes into play.

The term position limit refers to a limit on the number of contracts held in commodity derivatives by an individual or a specified class of market participants. This aims at preventing that individual market participants or groups become powerful enough to manipulate the market. Critics, however, fear that position limits reduce market liquidity and thus lead to increased price volatility and higher costs of hedging strategies.

While the EU Commission prefer implementation of position limits as optional, the Council and Parliament favour mandatory position limits in agricultural products, contrary to e.g. interest-rate derivatives. For different market participants there should be different reporting requirements that are based on the categories of the US futures market authority CFTC. Exemptions from adherence to the position limits can be approved only if positions are held solely for hedging of physical transactions, directly related to the commercial activity of non-financial entities. The issue of position limits has been controversial for a long time. The proposals of the Parliament and the EU Commission only relate to positions held on commodity exchanges, while the Council leaves this open in its formulation[4].

The level of position limits is to be negotiated in the so called Level-2-negotiations at EU level. This process has already been initiated by ESMA and an advisory group set up, consisting of representatives of affected market players, like agricultural associations and the food industry. A public consultation process is scheduled for late 2013.

Rather than the *principles*, the *details of implementation* will be key for how the regulation actually affects business in practice. There are still many open points, where the method of implementation greatly determines to what extent entrepreneurial activity is affected. These include the following questions:

- Is an EFP contract (Exchange of Futures for Physicals, "Prämiengeschäft"), e.g., a trade priced with discount or premium to the LIFFE-price, liable for clearing?
- Is it considered "speculation" if a position is opened to hedge for a certain quantity of wheat offered in a tender? At that point in time, no physical transaction has taken place, and possibly won't do so at all, if the offer within tender is rejected by the buy-side.
- What about "cross hedging" of a physical rye-transaction by taking a position in a related derivatives contract, e.g. wheat, at the commodity exchange?

In September 2013 the trilogies negotiations started on the basis of various proposals of the European Commission, the European Parliament and the European Economic and Financial Affairs Council ECOFIN. However, the positions were not in line. A compromise was intended for 2013 to avoid interference with the

---

4 An internal paper of the Federal Ministry of Finance published by Foodwatch (2013) shows that even within the trilogies parties the positions are still being readjusted: Previous positions of the Ministry of Finance have been significantly changed showing a stricter attitude towards regulation: treasury activities shall not be exempt from the regulations, position limits at the level of the ESMA should be set in a harmonized way and the limits shall apply to not only for individual market participants but aggregated for entire groups of actors.

European elections in 2014, but was not yet reached at the time of completion of this paper. It seems unlikely that any new regulation will come into force before 2015.

Amendment of the Market Abuse Directive MAD (MAD II)

A regulation and a directive target the prevention of insider trading and market manipulation. This also includes so-called "cornering", i.e., buying and hoarding a large share of a physical good on a specific market, simultaneously buying corresponding futures contracts and selling them at a profit after the price has been inflated by the artificial physical scarcity. This can make the prices explode, especially in a smaller market. Such abuse shall now be criminally prosecuted. This is an uncharted territory, and there exists only draft legislation. At this point, the interdependence of the different regulatory approaches arises. Indeed, while at the end of June 2013, a trilogies agreement on anti-abuse rules was achieved; further action on this topic requires results on MiFID as many important terms and concepts still await clear definition there. Again, the details will determine the effects. For example: Can information gained from own commercial activities - which are in part public and in part non-public - be considered insider information? And, if this were the case, would a corresponding hedging activity on the derivatives market be prohibited? This concerns for e.g. sugar companies that operate worldwide and from their commercial activities gain public as well as non-public information about the market (DÜLKS, 2013).

Regulation on the other side of the Atlantic

Position limits have existed in the USA for several decades, but they were mitigated by a number of revisions. Today, in many cases not the CFTC itself but only the commodity exchange sets the limits. The CFTC may request a more detailed surveillance of a market participant if it exceeds certain high position thresholds, in order to prevent market manipulation. However, in recent years the CFTC has hardened its stance, for example by not raising the limits further and even eliminating some exemptions for several financial market actors.

In the US, the Dodd-Frank Act of 2010 sets the regulation and supervision of the derivatives markets and transmits appropriate tasks to the competent authorities such as the CFTC and the SEC. In many areas on both sides of the Atlantic, the approaches pursue the same goals, namely the inclusion of the OTC market and increased transparency in general, and foresee such measures as clearing and reporting requirements as well as more security in non-cleared transactions. But there are also many differences, such as the following:

- Different to EMIR, for US non-financial companies there is no threshold for position size under which central clearing is unnecessary. Under Dodd-Frank there are only exceptions for so-called end-users with the direct aim to hedge a specific transaction risk.
- With reporting requirements EMIR allows more flexibility in reporting, e.g., not in real-time but at closing; the reporting obligations can be transmitted to a third party such as a bank. However, reporting requirements for NFC are stricter in the EU than in the US.
- Through Dodd-Frank, the CFTC has already got an assigned duty to set stricter position limits. However, the CFTC is faced with court claims, saying it has not adequately proved the necessity and appropriateness of the stricter position limits it administered in January 2012. At present, the legal problems are still not resolved (ÖFSE, 2013, pp. 6-7). Any new developments in the US with this regard will be followed in Europe with great interest.

Perspective of a financial transaction tax

Eleven EU countries, including Germany, have already agreed to introduce a financial transaction tax. In Germany, the financial transaction tax is also an issue in the current coalition talks between the CDU/CSU and SPD. A consensus has already been reached to further support and promote this project at the EU level. Details are still vague; however, a tax rate of about 0.1% on transactions in shares and securities, and 0.01% (or 0.005% for pension funds) on derivatives trading is discussed. Based on own simulations on the effects of financial transaction tax, the EU Commission expects the trade in shares/securities and in derivatives to be reduced by 15% and by 75%, respectively (EU Commission, 2013, p. 22). It considers this to be a "normalization", and sees no risk of liquidity shortage on the market: "…the Commission concluded in its analyses that the FTT (*financial transactions tax*) will not negatively affect the financial markets' efficiency. Nor will it result in the markets drying up and there being fewer possibilities for hedging risks, even if the number of transactions on the financial markets – and especially derivative bets – does shrink. Instead, the FTT will help to bring about a 'normalisation' of transactions on the financial markets…" (EU COMMISSION, 2013a, p. 5).

Conclusion

As of autumn 2013, the progress in formulation of new regulations on derivatives markets is still inconsistent between different countries and regions as well as different policy areas. The most advancements were achieved with respect to comprehensive reporting requirements. In the US and the EU, legal steps have been taken to implement central clearing, the obligation to trade on exchanges or exchange-like trading platforms, and stricter safety regulations for non-centrally cleared derivatives transactions. However, the legal processes as well as the technical implementation are still not completed.

It is questionable whether the policy is targeting the real problem. The discussion on the financialization of (agricultural) markets may be no more than a sideshow and distract from more important issues with regard to the objective of fighting world hunger. There is evidence suggesting that hunger is rather a result of country-specific problems and political discrimination. In addition, ad hoc trade measures may drastically increase price reactions caused by fundamental shortages, especially during periods of low stocks-to-use ratios (cf. SCHMITZ and MOLEVA, 2013).

The final design, implementation and any possible re-adjustments in the new regulations will determine whether the agricultural and food sectors will suffer under the new regulatory regime or whether they will still be able to manage their risks through tailor-made and affordable hedges.

List of References

BIS Bank for Internationale Settlement (2013): Derivatives Statistics, http://www.bis.org/ statistics/derstats.htm.

Bass, H.-H. (2013): Finanzspekulation und Nahrungsmittelpreise – Anmerkungen zum Stand der Forschung. Studie für foodwatch e. V. Materialien des Wissenschaftsschwerpunktes Globalisierung der Weltwirtschaft 42, Bremen.

Federal Ministry of Finance (2013): Fighting food speculation, June 6, Germany.

DAI Deutsches Aktieninstitut (2012): Risikomanagement mit Derivaten bei Unternehmen der Realwirtschaft - Verbreitung, Markttendenzen, Regulierungen - DAI-Kurzstudie 2/2012.

Dülks, N. (2013): EU Sugar Producers and Financial Market Regulation. Presentation to the Expert Group on agricultural commodity derivatives and spot markets. Brussels, October 3.

Etienne, X.L., Irwin, S.H. and Garcia, P. (2013): Bubbles in Food Commodity Markets: Four Decades of Evidence. Paper prepared for presentation at the Conference on "Understanding International Commodity Price Fluctuations" at the International Monetary Fund. Washington, D.C., March 20-21.

EU Commission (2013a): FTT – Non-technical answers to some questions on core features and potential effects, ec.europa.eu/taxation_customs/resources/documents/taxation/other_taxes /financial_sector/faq_en.pdf.

EU Commission (2013): Impact Assessment Accompanying the document 'Proposal for a Council Directive implementing enhanced cooperation in the area of financial transaction tax', Analysis of policy options and impacts. Brussels.

Foodwatch: Papier zu Änderung der Position des Finanzministeriums. Oktober 2013 http://www.foodwatch.org/de/informieren/agrarspekulation/aktuelle-nachrichten/ nahrungsmittelspekulation-kurswechsel-von-schaeuble/?sword_list%5B0%5D=schäuble, 01. 10.2013. Paper: http://www.foodwatch.org/uploads/media/MiFID_August_2013_DE_comments.pdf.

Guilleminot, B., Ohana, J.-J. and Ohana, S. (2013): The interaction of speculators and index investors in agricultural derivatives markets. Riskelia und ESCP Europe.

Kaya, O. (2013): Reforming OTC derivatives markets, DB Research Current Issues, Global Financial Markets.

ÖFSE (Österreichische Forschungsstiftung für Internationale Entwicklung) (2013): Re-regulation of commodity derivative markets – Critical assessment of current reform proposals in the EU and the US. Working Paper 45, Vienna.

Schmitz, P.M. and Moleva, P. (2013): Bestimmungsgründe für das Niveau und die Volatilität von Agrarrohstoffpreisen auf internationalen Märkten. Sind Biokraftstoffe verantwortlich für Preisschwankungen und den Hunger in der Welt? Justus-Liebig-Universität Giessen.

World Bank (2013): Global Economic Prospects, Commodity Markets Outlook, 2 July 2013.

Will, M.G., Prehn, S., Pies, I. and Glauben, T. (2012): Schadet oder nützt die Finanzspekulation mit Agrarrohstoffen? – Ein Literaturüberblick zum aktuellen Stand der empirischen Forschung. Diskussionpaper 26, Martin-Luther-Universität Halle-Wittenberg.

The Future Food Value Chain
Christian FISCHER

Abstract

This essay explains the supply and value chain concepts in the context of the global food system. Their origins and driving forces as well as recent trends and perspectives for future developments are discussed.

Introduction

As the world grows together, the way the production, distribution and consumption of goods and services are organised is evolving (HLGCAGI, 2009). The „relational view" of competitiveness is replacing the more traditional resource-based or market-based views which see economic success as a function of access to material resources or as a result of exercising market power (SPORLEDER et al., 2005; LATRUFFE, 2010). However, in a networked economy those organisations that form smart partnerships will have the competitive edge. By combining complementary core competencies, partnerships can benefit from synergies and collaborative advantages. These include traditional efficiency improvements through cost reductions and/or productivity gains and the co-creation of intangible assets resulting from collaborative innovation, leveraged mass communication and co-ordinated lobbying (KLAPWIJK, 2004).

A changing world

During the last decades, economic, technological, social and policy environments have structurally changed (GEREFFI et al., 2005). This is reflected in globalisation (i.e., widespread cross-border trade, investment and migration following market liberalisation and economic integration), urbanisation, the ubiquity of information and communication technologies, and ever better educated, informed and demanding workers and consumers. As a result, several types of "business systems" have emerged. These include geographically localised industry clusters (i.e., "Silicon Valley" type of technology parks), horizontal production networks (e.g., globally dispersed car manufacturer suppliers), and vertical, producer-to-consumer "chains" (KLAPWIJK, 2004). Business systems are competitive structures because they allow collaborating organisations to exploit economies of size (scale and scope) while maintaining strategic flexibility, operational agility and entrepreneurial initiative.

In the agrifood sector, the concept of value chains has been heavily promoted and applied as a means of fostering agricultural development, and in particular of linking small-scale farmers to markets. Leading global institutions such as the World Bank, the Food and Agricultural Organisation of the United Nations (FAO) and the United Nations Industrial Development Organisation (UNIDO) have issued studies on agrifood value chains in the recent past (HUMPHREY and MEMEDOVIC, 2006; FIAS, 2007; UNIDO, 2011, FAO, 2013). Yet value chains are also a key topic for the developed world"s food multinationals. For example, the Global Commerce Initiative, with members such as Kraft, Unilever, Nestlé, Procter & Gamble, Wal-Mart, Carrefour, Royal Ahold etc., has established a value chain project (GCI, 2006; GCI, 2008; CGF, 2013). The mission is to better align consumer needs and wants with producer and distributor profit requirements and their technological and organisational capabilities, while minimising resource waste and other negative environmental impacts.

Value capturing and creation

There is considerable confusion about the chain terminology. Value, supply or agrifood chains mean different things to different people. Supply chain management has become a recognised, multi-disciplinary academic field and an established industry practice. Two views on the subject are common: upstream, procurement and supply management, and downstream, distribution and demand management. Both views are mostly concerned with managing material and information flows (VAN DER VORST et al., 2007). The value chain approach adds the financial dimension.

While in theory financial remuneration along the chain should be proportional to employed resources (labour, skills, capital), in practice „value capturing" is often a result of unbalanced power constellations within the chain. This situation raises questions of distribution fairness and is reflected in the existence of global campaigns such as "fair trade". Another approach is to define „value" more widely, and to incorporate externalities (BARBER, 2008). In particular for food products, "credence" attributes related to production (child labour, animal welfare, chemical use etc.) and distribution issues (food miles, fair trade) have become more and more important purchasing motives for consumers (CAPGEMINI, 2007). Smart marketing, addressing these new consumer concerns, can add a lot of value to commodities and understanding this process is part of value chain analysis.

More generally, the „Triple P" approach of value creation („people", the „planet" and „prosperity" or „profit") has become more widely accepted. Corporate responsibility no longer stops at shareholders and in order to stay competitive, organisations need to find ways to be simultaneously financially viable, socially responsible and environmentally sustainable (VAN DER VORST et al., 2007; LATRUFFE, 2010). In particular the latter is of high current concern (SMITH, 2008; GODFRAY et al., 2010; ECONOMIST, 2011). As a consequence, the chain concept that traditionally covered the way "from producer to consumer" has now been extended to "from nature to nature", including pre-production links (e.g., natural resource-related research) and post-consumption issues (e.g., waste disposal and recycling). In the future, „value cycles" may be replacing „chains" as life cycle assessments complement value chain concepts.

Collaboration along the chain

Creating value from business partnerships (i.e., a group of knowledge-sharing but independent enterprises aligned in a non-hierarchical way) is as much about technology as organisation (UNIDO, 2011; MALIK et al., 2011). Establishing fast and comprehensive data exchange is at the core of optimising warehousing and transportation and thus to improve logistics. This includes sophisticated software as well as leading-edge physical devices (e.g., RFID tags). But turning an efficient supply chain into an effective and sustainable value chain also involves enhanced, strategic collaboration among chain partners.

Traditionally, economic organisation was a question of "make" (yourself) or "buy" (from others if you were not able to produce a part etc.). Now the paradigm is to "ally" – i.e., to make together. However, there are many practical organisational issues involved. These include identifying suitable partners, aligning strategies, deciding on the type of contracts to be used, defining the exact nature of collaboration, establishing and nurturing inter-enterprise relationships, dealing with conflicts, and building and maintaining trust among partners (VAN ROEKEL et al., 2004; SPORLEDER et al., 2005; VAN DER VORST et al., 2007).

Given that getting organisations to work together is easier said than done, governments in many countries have set up chain facilitators and business network brokers, or have established partner search platforms. There are also capability-building schemes to train managers to be effective within a business partnership (VAN ROEKEL et al., 2004). Being part of an inter-enterprise collaboration arrangement is not in itself a panacea for business success. If its establishment and continuous operation are not properly managed, the partnership will fail. Reasons for failure often come down to not actively addressing differences between partners, be they strategic (i.e., different goals), operational (i.e., different production and/or distribution approaches) or cultural (different ways of dealing with staff and/or customers) (BITITCI et al., 2009).

The way ahead

The future food value chain will have different formats and facets. That is, there will be many successful models existing simultaneously (FT, 2010; FT, 2012; FT, 2013; KING et al., 2010). There will be industrial food manufacturers that formulate products specifically to individual consumer requirements (mass-customisation), supported by highly fine-tuned logistics. There will be the small-scale rural farmer who sells on-farm local meat from extensively reared livestock, slaughtered on the premises, at attractive price premiums to environmentally concerned city dwellers. There will be metropolitan agriculture with employed „biological production specialists" growing efficiently high-value speciality crops on roof tops and in integrated glass house production systems in and nearby urban areas. There will be supermarkets venturing into food service, offering non-industrial, hand-crafted ready-made meals on a large-scale basis.

As our definitions of „value" shift, so will the ways to deliver it. What stays is the need for knowledge generation, update and transfer. That is, the need to identify, analyse and document new – often private – value creation mechanisms and to turn them into public knowledge so that they can be multiplied and the world as a whole can advance. Even if research is only a first step in a long and complex innovation value chain, without it, social progress will not happen.

List of References

Barber, E. (2008): How to measure the "value" in value chains, International Journal of Physical Distribution & Logistics Management, 38 (9), 685-698, DOI: 10.1108/09600030810925971.

Bititci, U., Butler, P., Cahill, W. and Kearney, D. (2009): Collaboration - A key competence for competing in the 21st century. Supply Chain Perspectives, 10 (1), 24-33, Irish National Institute for Transport and Logistics.

Capgemini (2007): Future Consumer: How Shopper Needs and Behaviour Will Impact Tomorrow"s Value Chain, http://apps.us.capgemini.com/FutureConsumer.

CGF (2013) (The Consumer Goods Forum) Future Value Chain 2022 – Industry Initiatives Address Challenges of the Digital World and the Fight for Resources, www.futurevaluechain.com.

Economist, The (2011): The 9-billion People Question – A Special Report on Feeding the World, February 26, www.economist.com/printedition/specialreports.

FAO (2013) (Food and Agriculture Organization of the United Nations): The State of Food and Agriculture 2013: Food Systems for Better Nutrition, Rome, www.fao.org/ publications.

FIAS (2007) (The Foreign Investment Advisory Service of the World Bank Group): Moving Toward Competitiveness: A Value Chain Approach, World Bank Group, Washington, DC, www.worldbank.org/publications.

FT (2012) (Financial Times): The Future of the Food Industry, FT Special Report, November 21, www.ft.com/reports.

FT (2013) (Financial Times): The Future of the Food Industry, FT Special Report, November 20, www.ft.com/reports.

FT (2010) (Financial Times): World Food, FT Special Report, October 15, www.ft.com/ reports.

GCI (2006) (Global Commerce Initiative): Capgemini, 2016 Future Value Chain, www. futuresupplychain.com/downloads.

GCI (2008) (Global Commerce Initiative): Capgemini, Future Supply Chain 2016: Serving Customers in a Sustainable Way, www.futuresupplychain.com/downloads.

Gereffi, G., Humphrey, J. and Sturgeon, T. (2005): The governance of global value chains, Review of International Political Economy, 12 (1), 78-104, DOI: 10.1080/096922905 00049805.

Godfray, C., Crute, I., Haddad, L., Lawrence, D., Muir, J., Nisbett, N., Pretty, J., Robinson, S., Toulmin, C. and Whiteley, R. (2010): The future of the global food system, Philosophical Transactions of the Royal Society B, 365, 2769-2777, DOI: 10.1098/rstb. 2010.0180.

HLGCAGI (2009) (High Level Group on the Competitiveness of the Agro-Food Industry): Report on the Competitiveness of the European Agro-Food Industry, European Commission, Enterprise and Industry Directorate General, Food Industry Unit, http://ec.europa. eu/enterprise/sectors/food/competitiveness/high-level-group.

Humphrey, J. and Memedovic, O. (2006): Global Value Chains in the Agrifood Sector. United Nations Industrial Development Organisation (UNIDO), Vienna, Austria, www.unido. org/publications.

King, R., Hand, M., DiGiacomo, G., Clancy, K., Gomez, M, Hardesty, S., Lev, L. and McLaughlin, E. (2010): Comparing the Structure, Size, and Performance of Local and Mainstream Food Supply Chains, ERR-99, US Department of Agriculture, Economic Research Service, www.ers.usda.gov/briefing.

Klapwijk, P. (2004): Economics Reconfigured – A Vision of Tomorrow''s Value Chains, Summarised inaugural speech in acceptance of the Chair of Supply Chain Economics at the Universiteit Nyenrode, www.nyenrode.nl/download/lectures/klapwijk.pdf.

Latruffe, L. (2010): Competitiveness, Productivity and Efficiency in the Agricultural and Agri-Food Sectors, OECD Food, Agriculture and Fisheries Working Papers, No. 30, OECD Publishing, DOI:10.1787/5km91nkdt6d6-en.

Malik, Y., Niemeyer, A. and Ruwadi, B. (2011): Building the supply chain of the future, McKinsey Quarterly, No. 1, 1-10.

Roekel, van J., Kopicki, R., Broekmans, C. and Boselie, D. (2004): Building agri-supply chains: issues and guidelines, in Giovannucci, Daniele (ed.): A Guide to Developing Agricultural Markets and Agro-enterprises, http://go.worldbank.org/ 1DBLU3WAQ0.

Smith, G. (2008): Developing sustainable food supply chains, Philosophical Transactions of the Royal Society B, 363, 849-861, DOI:10.1098/rstb.2007.2187.

Sporleder, T., Cullman C. and Bolling, D. (2005): Transitioning from transaction-based markets to alliance-based supply chains: implications for firms, Choices: The Magazine of Food, Farm, and Resource Issues, 20 (4), 275-280.

UNIDO (2011) (United Nations Industrial Development Organization): Industrial Value Chain Diagnostics: An Integrated Tool, Vienna, Austria, www.unido.org/publications.

Vorst, van der J., Silva, da C. and Trienekens, J. (2007): Agro-industrial Supply Chain Management: Concepts and Applications, Agricultural management, marketing and finance occasional paper, No. 17, Food and Agriculture Organization of the United Nations (FAO), Rome, www.fao.org/publications.

Impacts of an EU-USA-Free Trade Agreement on Developing Countries

Martina BROCKMEIER, Tanja ENGELBERT and Janine PELIKAN

Abstract

A successful conclusion of the Transatlantic Trade and Investment Partnership (TTIP) between the European Union (EU) and the United States (US) will create one of the biggest Free Trade Areas (FTAs) world-wide between two already very important trading blocks. The analysis in this paper is based on an integrated econometric-CGE (Computable General Equilibrium) analysis, where econometrically estimated non-tariff barriers (NTBs) are used in conjunction with the GTAP (Global Trade Analysis Project) model. We estimate NTBs for a disaggregated food and agricultural sector comprising 16 primary agricultural and processed food sectors. The results show that the elimination of NTBs between the EU and US is of substantially higher importance for TTIP than the abolishment of tariffs. The EU and the US will experience a noteworthy increase of their welfare and GDP, whereas we notice a negative development for these variables in third countries. However, a clear trend in the results for the food and agricultural sectors of the two partners of the TTIP cannot be observed. None of the TTIP countries can consistently increase their exports of food and agricultural products, nor does one show a consistent decrease in imports of these sectors. On the contrary, it seems to be clear that third countries are more likely to gain from the TTIP if they adapt to the standard set by the EU-US FTA and take advantage of the resulting spill-over effect.

Introduction

In June 2013, the European Union (EU) and the United States (US) started negotiating a Free Trade Area (FTA). A successful conclusion of this Transatlantic Trade and Investment Partnership (TTIP) will create one of the biggest FTAs world-wide. First attempts to achieve this FTA were already initiated in the 1990s and were renewed in 2006/2007. Recently, the negotiations have been resumed, and after the High Level Working Group on Jobs and Growth approved deep trade integration between the EU and the US in February 2013, the conclusion of the negotiations for the TTIP has come within reach.

Decisive for the opening of the trade talks between the EU and the US have been the lack of progress in the Doha Round of the WTO negotiations, but also the increasing erosion of competitiveness of industrial countries and crisis-related necessity for structural reforms to strengthen the economy. By opening markets and increasing investment opportunities, the two giants of world trade are trying to drive growth and create jobs to boost their economies and keep up with other global players, particularly with the fast-growing emerging countries. Tariffs have already been reduced between the EU and the US over the last decades due to the WTO negotiations, although this liberalisation has predominantly taken place in the non-agricultural sector. The EU and the US therefore expect the highest gain to come from the harmonising of the EU and US technical standards and other non-tariff barriers (NTBs). An agreement on NTBs between the EU and the US might also provide the basis for common global standards. The related spill-over effects in third countries could contribute to reducing the expected trade-diverting effect of the TTIP.

Up until now only very few studies are available which analyse the impact of the Transatlantic FTA. Their results confirm that both partners of this FTA would gain from the successful conclusion of the negotiations and that the main benefit would result from the willingness to make EU and US trade rules and regulations compatible (e.g., CEPR, 2013; FELBERMAYR et al., 2013). The harmonisation of NTBs is, however, discussed controversially in public. Particularly important in this discussion are the NTBs of the food and agricultural sector. Here, consumer organisations and other non-governmental organisations in Europe already fear that the TTIP will lower the safety and environmental standards in the EU, but also the US. However, none of the available studies on the impacts of the Transatlantic FTA between the EU and the US takes a disaggregated food and agricultural sector into account, so that a detailed analysis of the harmonisation of the EU and US trade rules and regulations in the food and agricultural sector is not available, yet.

In light of the above, we analyse the effect of the TTIP taking a detailed food and agricultural sector into account. The paper is organized as follows. The second chapter provides a brief overview of the work of P. Michael SCHMITZ on how the common agricultural trade policy of the EU (CAP) and European Trade Agreements influenced developing countries, and on how he thereafter motivated the authors of this article to utilize different tools like computable general equilibrium (CGE) models to analyse these issues. For reasons of space, we thereby disregard the numerous publications of other scholars at the Institute of Ulrich KOESTER and of P. Michael SCHMITZ who also deal with the effect of the CAP on developing countries utilizing non-CGE related methods. The third Chapter provides an overview of the trade integration and protection structure of the EU and the US. Thereafter, we discuss the method to econometrically estimate the trade costs of NTBs relevant in the bilateral trade flows between the EU and the US. In a second step these estimates are integrated into the Global Trade Analysis Project (GTAP) model to quantify the economy-wide effects of the TTIP. The fourth Chapter summarises the main findings of the paper.

Analysis of Trade, Preference and WTO Agreements with CGE Models: Schmitz and his Scholars

How does EU agricultural trade policy influence developing countries? This question is as old as the EC/EU itself. Researchers have analysed this question in numerous studies with the help of many different methods and, depending on the point in time, have taken different economic and political environments into account.

From its early beginnings the EC initiated trade agreements like the preferential market access agreement arising from former colonial relationships with the African, Caribbean, and Pacific Group of States (ACP). Due to EC accessions of further member countries, additional preferences were granted for selected developing countries like import quotas on sugar or beef. Research on EC trade policy concentrated at that time on these non-reciprocal trade preferences. On the one hand, the EC trade policy instruments like export subsidies depressed and destabilized world market prices and thus especially harmed exporting developing countries. On the other hand, some developing countries benefited from their preferential market access to the EC.

SCHMITZ has always shown a particular interest in this issue. He analysed welfare effects of EU trade preferences on developing countries by applying partial equilibrium approaches like in SCHMITZ and KOESTER (1981), KOESTER and SCHMITZ (1982). He found, e.g., for the sugar market that negative welfare effects on preference receiving countries due to lower world market prices are overcompensated by positive transfer effects due to higher EU prices. However, the authors also found several negative by-product distortions that come with these transfer effects. Additionally, SCHMITZ (1982, 1984) analysed instabilities on agricultural world markets measured by the coefficient of variation and applied nonlinear functions with multiplicative disturbance terms. Here, he found that welfare gains due to EU trade liberalization are underestimated if only the price level is taken into account. Additional welfare gains arise due to price stabilisation for exporting countries. For countries that receive preferential market access to the highly protected EU market, a substantial stabilization effect occurs which can be greater than the gain from complete trade liberalization. SCHMITZ suspected that this result might give an incentive for preference receiving countries to focus on bilateral negotiations with the EU rather than to negotiate at the multilateral level.

Indeed, the following decades were characterized by long and cumbersome multilateral trade talks of the GATT/WTO, while the EU aimed to include the up to this point highest number of member countries with its Central and Eastern European Enlargement. SCHMITZ was one of the first researchers in Germany who recognized that the effects of both issues are capably analysed with the help of CGE models. Accordingly, he invited internationally renowned researchers in this area to work at his institute. SCHMITZ also organized high profile conferences where bilateral and multilateral trade negotiations and their impacts on developing countries were discussed. Among other advanced quantitative methods, CGE models and particularly the GTAP model were used in the analysis. In so doing, SCHMITZ strongly motived his scholars to employ CGE models in their research work.

The EU-Enlargement and the WTO negotiation provided many opportunities to apply CGE models. The Agreement on Agriculture of the Uruguay Round (1986–1994) required countries to reduce agricultural support and protection in three areas: market access, domestic support and export subsidies. The Doha Round (2001-2013) put a particular focus on developing countries and aimed to further liberalize agricultural markets in all three pillars. Preference erosion for developing countries was identified as one of several other reasons for the difficult consensus in multilateral negotiations (BROCKMEIER and PELIKAN, 2007). Thus, preference erosion due to more trade liberalization, but also additional preferences for Least Developed Countries (LDCs) were on the agenda. The EU trade policy was more and more embedded in this multilateral framework and research addressed the following questions: How do tariff reduction formulas negotiated in the WTO influence developing countries? What are the trade and welfare effects on developing countries due to preference erosion? Scholars of SCHMITZ analysed the welfare effects of multilateral trade liberalization with the GTAP model. This model was extended with EU specific features like the common EU budget, milk quotas and the decoupled direct payments (BROCKMEIER, 2003) or a specific tariff module that includes data on bound and applied tariff rates and enables tariff cuts at the HS 6 tariff line to be calculated based on tiered formulas (PELIKAN, 2009). They found that developing countries could gain from multilateral trade reforms. This result is mainly due to the expansion of market access. In contrast, the abolishment of export subsidies in developed countries deteriorates the term of trade effect and results in negative welfare impacts for developing countries who are net food importers.[1] LDCs, in contrast, experience the highest welfare gains if tariff cuts of developed countries are low in agriculture. This result mainly comes from the erosion of preferences that the EU grants the LDCs within the framework of the Everything but Arms-Initiative (BROCKMEIER and PELIKAN, 2008a, 2008b).

---

1   The stabilization effect analysed intensively by earlier work of SCHMITZ was not quantified in these analyses.

Many countries entered into bilateral and regional trade negotiations in parallel to the difficult and lengthy WTO negotiations. Figure 1 shows the expansion of trade agreements reported to the WTO during the last decades. Here, it becomes apparent that trade liberalization in form of free trade agreements is in vogue.

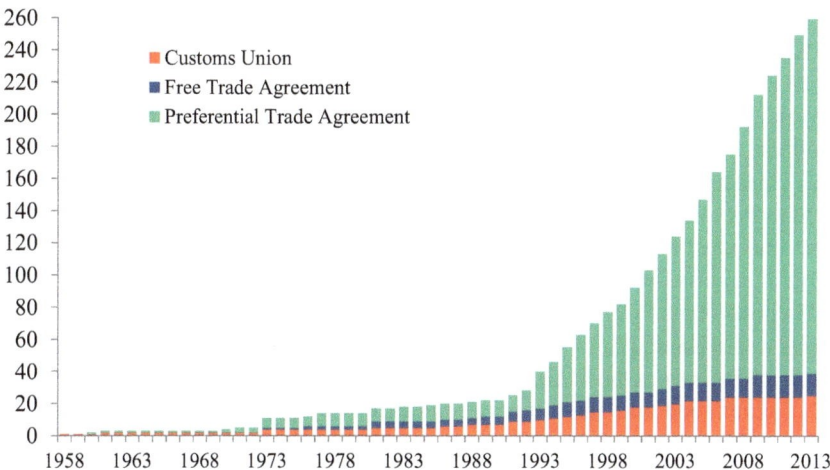

Fig. 1: *Number of Trade Agreements Reported to the WTO by Year*
Source: WTO, 2013

The exceptional WTO-rule for the non-reciprocal ACP-agreement of the EU expired in 2007 and a replacement started with reciprocal Economic Partnership Agreements. Additionally, the EU initiated many bilateral trade talks and already concluded free trade agreements with Peru, South Korea, Chile, Columbia and, most recently, Canada.

While the policy focus shifted from multilateral to free trade negotiations, the methods to analyse the underlying trade agreement also changed. In the past, mainly econometrics, partial or general equilibrium models have been separately applied. Today, the combination of these approaches allows for more detailed impact assessments of various trade policies, e.g., the explicit consideration of NTBs in general equilibrium analysis. This is the basis for the subsequent empirical analysis. With a combination of results from gravity modelling and the GTAP model, this paper contributes to the literature by analysing the FTA between the EU and the US and its impacts on developing countries. In contrast to the discussed preferential or multilateral agreements, developing countries themselves have no influence on the negotiations, but will be affected due to trade diversion or creation.

The Transatlantic FTA between the EU und the US: An integrated Econometric-CGE Analysis

The EU and the US are already linked through intensive trade relations. Both partners of the TTIP account for over half of the worldwide GDP and almost one third of world trade. The EU and the US are also important trade partners for other countries. A successful conclusion of the Transatlantic FTA would therefore greatly reshape the global trade structure (EU Commission, 2014).

In terms of agro-food trade, the US is the most important export partner and the second most important import partner for the EU in 2012. 13.1% of total agricultural exports from the EU go to the US market and 8.2% of the total agricultural imports come from the US. Agricultural products amount to 6.8% of the total exports to the US and 5.7% of the total imports from the US. The most important export products from the EU to the US are alcoholic and non-alcoholic beverages (52%), other food and agricultural products (40%) and dairy (4%). Other food products (51%), tropical fruits and spices (16%), alcoholic beverages (13%) and soya beans (11%) have the highest proportion of the food and agricultural imports of the EU from the US of the total agricultural imports (EU Commission, 2013).

The average tariff of the EU is about 1.8% and the US average tariff is about 0.97%; whereas food and agriculture tariffs (EU: 8.41%; US: 3.38%) are generally higher than in the non-agricultural sectors (EU: 2.03%; US: 1.27%). The EU protects its markets more than the US. Except for the sugar sector, all import tariffs are higher in the EU compared to the import tariffs of the US. The EU raises its highest tariffs on US exports of beef (42%), dairy (40%), pork and poultry (38%), and rice (24%). In contrast, the US raises its highest tariffs on EU exports of dairy (19%), sugar (15%), other crops (7%) and rice (6%) (GTAP Database, version 8).

As stated above, gains from a Transatlantic FTA will mainly come from the reduction of NTBs and harmonization of technical regulations, standards and approval procedures. Thereby, it is stressed by both parties that the lowering or eliminating of standards and technical regulations, or limiting the autonomy to set regulations is not a matter of debate. Accordingly, in the third round of negotiations in December 2013, it was emphasised that within the TTIP the high standards of consumer, environment, health and labour protection will be upheld. However, while keeping a high level of standards it is still possible to avoid unnecessary transfer and administrative costs. This can be achieved, for example, when both parties recognize similarities in approval procedures, adjust the wording of provisions to comply with each other's existing rules or cooperate more closely in the future when new rules are designed (EU Commission, 2014).

Although researchers acknowledge the important role of NTBs in the analysis of the TTIP, it is unclear how to quantify the related costs. There are several methods to capture the enormous variety of non-tariff trade measures in one uniform metric. In the following, the standard procedure of gravity modelling is used to estimate ad valorem tariff equivalents (AVEs) of non-tariff trade barriers. In the second step the AVEs are implemented into a CGE application to derive economy-wide results. Thereby the main focus is on the effects of the TTIP on the food and agricultural sector and on developing countries taking spill-over effects into account.

Estimation of Non-Tariff Barriers using the Gravity Model

Theoretical and Empirical Framework

The estimation of trade costs of NTBs is based on a cross-sectional analysis for 2007 using the gravity model. The gravity model has become the standard model to empirically measure expected bilateral trade using economy size and an additional set of control variables.[2] The model is frequently used to estimate NTBs because of the sound theoretical justification and simple and flexible application. The following analysis adopts the structural gravity-like equation developed by ANDERSON and VAN WINCOOP (2003, 2004). Their model takes the following form:

$$X_{ij} = \frac{Y_i Y_j}{Y_w} \left(\frac{t_{ij}}{P_j \Pi_i} \right)^{1-\sigma} \qquad (1)$$

Where the exports from country i to country j (X_{ij}) depend on exporter production (Y_i) and importer consumption (Y_j) relative to the global output (Y_w). The main interest is on bilateral trade barriers (t_{ij}). They reduce trade as determined by the elasticity of substitution (σ). Average trade barriers are represented by outward trade barriers (Π_i) of country i and inward trade barriers (P_j) of country j which are known as multilateral resistance variables (ANDERSON and VAN WINCOOP, 2003). The larger these barriers, the more bilateral trade between

---

2   See ANDERSON (2011) and HEAD and MAYER (2013) for a detailed review on gravity models.

country i and j occurs. Unobservable bilateral trade frictions are proxied empirically with observed trade costs factors; namely with natural and policy-based variables. Bilateral distance and geographical dummy variables are applied and assumed to be equal to one if regions i and j are both landlocked, are both islands or share a common land border. Also included are cultural dummy variables which are equal to one if regions i and j share a common colonial relationship and language. Furthermore, the log of one plus the ad-valorem tariff imposed by region j on imports from i is used to factor in the trade policy of the country. Finally, three regional dummy variables are included that are equal to one if i and j are both members of the EU or NAFTA, respectively, or if the dependent variable measures trade between Australia and New Zealand. The bilateral cost function reads as follows where ln indicates the logarithm of a variable:

$$t_{ij}^{1-\sigma} = exp\begin{pmatrix} \beta_1 \ln Dis\tan ce_{ij} + \beta_2 Landlocked_{ij} + \beta_3 Contiguity_{ij} + \beta_4 Island_{ij} \\ +\beta_5 Language_{ij} + \beta_6 Colonizer_{ij} \\ +\beta_8 \ln Tariff_{ij} + \beta_9 EU_{ij} + \beta_{10} NAFTA_{ij} + \beta_{11} AUSNZL_{ij} \end{pmatrix} \quad (2)$$

Substituting Equation (2) in (1), the multiplicative form is obtained

$$X_{ij} = exp\left(\delta_1 EU_{ij} + Z_{ij}'\beta + \alpha_i + \alpha_j\right) \quad (3)$$

Where Z_{ij} = (1; $Distance_{ij}$; $Contiguity_{ij}$, ...) is a vector which contains a constant and all variables promoting or hindering trade except the dummy variable for EU. Furthermore, $\beta = (\beta_0, \beta_1, \beta_2, ...)$ is a vector of the coefficients applying to the elements in Z_{ij}. The coefficient δ_1 gives the trade preferential effect existing in the EU. It is an implicit measure of non-tariff protection and is the key element in the following econometric analysis. Average trade barriers are controlled by specifying importer and exporter fixed effects (e.g., CHEN, 2004; FEENSTRA, 2002; OLPER and RAIMONDI, 2008; PHILIPPIDIS et al., 2013; WINCHESTER, 2009). Thus, dummies are included that are connected to each exporter's and importer's trade entity. As such, the country dummies control not only for multilateral resistance but also for unobserved country-specific factors. Accordingly, $\alpha_i = lny_i - (1-\sigma)ln\Pi_i$ is the fixed effect of the exporting country and $\alpha_j = lny_j - (1-\sigma)lnP_j$ is the fixed effect of the importing country. The constant β_0 is equal to $(-lny_w)$. Finally, an error term (ε_{ij}) is added to obtain an estimable equation. Equation (3) is applied to 16 agro-food sectors, 10 manufacturing and nine services sectors (see

Table A1 in the Appendix).³ The most important parameter to estimate is the coefficient of the EU membership dummy as a benchmark for the level of non-tariff trade measures. By using the EU-specific dummy variable it is possible to identify policy instruments that aim to harmonize standards and regulations within a specific region (CHEN and NOVY, 2012). The rationale to take the EU as the benchmark is based on the assumption that the overall level of regulatory divergence is the lowest among EU member countries. The specification allows the average partial effects of the most integrated trade relations to be identified. Using this method is an advantage for identifying all existing NTB-induced trade costs at the sector level and reaching a summary measure in a consistent manner.

Data and estimation strategy

Data on bilateral exports and tariffs are sourced from the most recent version 8 of the GTAP database. The GTAP database offers information about 134 regions and 57 sectors. The number of regions is reduced to 114 by omitting composite regions; the number of sectors is also reduced to the 16 food and agricultural sectors, 10 manufacturing sectors and nine services sectors. The regression analysis includes 12,882 observations for each sector. The information on distance, landlocked status, contiguity, common language and colonial relationships comes from the Centre D'Etudes Prospectives et D'Informations Internationales.⁴

Typically the gravity model is estimated in the form of the log-linear specification using ordinary least squares (OLS). One serious challenge that arises when applying OLS in this framework is the presence of zero trade flows. The dataset includes 12.2% of agricultural export flows, 17.8% of manufacturing exports and 8.9% of services exports which are equal to zero. Furthermore, OLS estimates are biased due to heteroskedasticity. Using the Poisson pseudo-maximum likelihood estimator it is possible to account properly for zero observations and to consider heteroskedasticity, making it favourable in gravity modelling.⁵

---

3   In estimating NTBs induced trade costs, we only use 16 of 20 food and agricultural sectors used in the simulation exercise omitting the generally non-tradable sectors paddy rice, sugar cane and beets, raw milk and wool.
4   See http://www.cepii.fr/CEPII/en/bdd_modele/bdd.asp.
5   See a detailed review on the performance of the Poisson pseudo-maximum likelihood estimator in SANTOS SILVA and TENREYRO (2006, 2011).

Estimation Results

Table 1 shows key results in the disaggregated agro-food sectors[6]. The magnitude of the EU estimate is the highest in the sugar sector. The effect of both countries being EU members is equivalent to an increase of trade in sugar by 2015.9%. Also very high effects are estimated for the sectors cattle meat (1197.7%), other meat (886.1%), dairy (759.9%) and processed rice (681.9%). EU membership has a relatively low impact on trade in the sectors such as beverages and tobacco (37.6%), cattle (44.1%) and other food (85.5%).

According to the theoretical specification of the gravity model, the overall multilateral effect of NTBs for each country is captured by its country fixed effect. To compute bilateral non-tariff trade costs, the quantitative effect of the EU-specific dummy considering a given level of NTBs multilaterally was utilized. It is assumed that the EU dummy reveals those gains reached through a reduction of NTBs induced trade costs. The following formula is used to calculate the costs of non-tariff trade barriers:

$$AVE_{ij}^k = exp\left(\frac{\delta_l}{\sigma_k - 1}\right) - 1 \quad (4)$$

Where AVE^k_{ij} is the trade costs in form of AVE of NTBs between i and j in sector k, δ_1 is applied to the EU membership dummy and σ_k is the elasticity of substitution in sector k. The GTAP database serves as source for the elasticity of substitution for the disaggregated sectors.

Table 1 also reports the AVEs of NTBs in the disaggregated agro-food sectors using the parameter ESUBM from the GTAP database. ESUBM is the elasticity of substitution among sources of imports in the Armington aggregation structure for all agents in all regions. The AVEs of the NTBs in the disaggregated food and agricultural sectors presented in this paper are, in most cases, reasonably similar to or lower than those given in the literature (PHILIPPIDIS and SANJUÁN, 2006, 2007; WINCHESTER, 2009).

---

[6] Econometric results in the aggregated sectors are reported in Table A2 in the Appendix. Detailed econometric results for the disaggregated agricultural and non-agricultural sectors are available upon request from the authors.

Simulations with the GTAP Model

GTAP Model and Data

The CGE simulations in this paper utilize GTAP, a comparative static multi-region general equilibrium model. The framework of the standard GTAP model is well documented in HERTEL (1997) and is available on the Internet.[7]

FRANCOIS (1999, 2001) developed an approach in which NTBs are modelled as iceberg or dead-weight costs and used this method to study the Doha Round of the WTO negotiations. This approach has been extended by HERTEL et al. (2001a, 2001b), who also aimed to integrate NTBs into GTAP modelling, treating NTBs as unobserved trade costs that are not explicitly covered by the GTAP database.

Tab. 1: *Key Results in the Disaggregated Agro-food Sectors*

	EU			Elasticity of Substitution	AVE %
	Coefficient	Std. Error	Effect %		
Wheat	2.0338***	(0.3122)	664.34	8.9	29.36
Cereal grain	0.6783	(0.3711)	97.06	2.6	52.80
Fruit and vegetable	1.2955***	(0.2113)	265.30	3.7	61.58
Oil seeds	0.9915**	(0.3622)	169.52	4.9	28.95
Plant-based fibres	1.5729***	(0.3690)	382.05	5	48.17
Other crops	0.6307*	(0.3035)	87.89	6.5	12.15
Cattle	0.3652	(0.4553)	44.08	4	12.95
Other animal products	0.6552*	(0.2658)	92.55	2.6	50.61
Vegetable oils and fats	0.8871**	(0.3275)	142.80	6.6	17.16
Processed rice	2.0566***	(0.5023)	681.92	5.2	63.18
Sugar	3.0521***	(0.4693)	2015.89	5.4	100.10
Other food products	0.6181***	(0.1205)	85.54	4	22.88
Beverages and tobacco	0.3192	(0.2359)	37.60	2.3	27.83
Cattle meat	2.5631***	(0.3831)	1197.65	7.7	46.60
Other meat	2.2893***	(0.2858)	886.81	8.8	34.11
Dairy	2.1517***	(0.1890)	759.94	7.3	40.71

Notes: Robust standard errors in parentheses. *, **, and *** indicate significance at 5, 1 and 0.1% levels, respectively

Source: Own calculation

---

7 See https://www.gtap.org

The authors introduce an additional effective import price that is a function of the observed import price and an exogenous unobserved technical coefficient (HERTEL et al., 2001a, p. 13):

$$pms^*_{irs} = pms_{irs} - ams_{irs} \qquad (5)$$

where pms^*_{irs} percentage change of the effective import price of i supplied from region r to region s
 pms_{irs} percentage change of the import price for i supplied from region r to region s
 ams_{irs} percentage change in the effective price of i from region r in region s due to changes in unobserved trade costs

The removal of trade costs from a particular exporter is reflected in an increase in ams_{irs} assuming that ams_{irs} is equal to one in the initial equilibrium. The effective import price of good i exported from region r to region s falls and thereby mirrors a reduction in real resource costs. This approach to model the change in NTBs as a reduction in trade costs draws on the iceberg transport cost theory that was originally introduced by SAMUELSON (1954).

An increase in ams_{irs} and the corresponding efficiency enhancement furthermore implies that the effective imported quantity of good i from region r to region s is increased, which leads to the following import demand and composite import price equation (HERTEL et al., 2001a, p. 13):

$$qxs_{irs} = -ams_{irs} + qim_{is} - \sigma^i \cdot (pms_{irs} - ams_{irs} - pim_{is}) \qquad (6)$$

$$pim_{is} = \sum_r \theta_{irs} \cdot (pms_{irs} - ams_{irs}) \qquad (7)$$

where qxs_{irs} percentage change in bilateral exports of i supplied from region r to region s
 qim_{is} percentage change in the total imports of i into region s
 pim_{is} percentage change in the import price of i in region s
 σ^i the elasticity of substitution among imports of i
 θ_{irs} the share of imports of i from region r to region s at market prices

According to equations (6) and (7), an increase in ams_{irs} implies that imports of i from region r to region s are more competitive and lead to the substitution of imports from other regions.

In addition, NTBs also generate protection effects that might be captured via import tariffs. ANDRIAMANANJARA et al. (2003, 2004) and FUGAZZA and MAUR (2008) offer a thorough study of the impact of NTBs in regional and global CGE models comparing the iceberg cost approach and the approach that involves capturing NTBs via import tariffs. Effects of NTBs are measured by the price wedges between domestic and world prices, when NTBs are modelled with the help of import tariffs. This import-tariff approach to represent NTBs creates a rent that is associated with the NTBs and is captured by the importer. Modelling NTBs with the help of the iceberg cost approach is also referred to as the "sand in the wheels" of trade or the "efficiency approach" by the authors. In the iceberg cost approach, it is thus assumed that NTBs are efficiency losses rather than rent-creating mechanisms, and as aforementioned, by using import-augmenting technology shocks, real resource cost raising effect of NTBs are abolished. The results obtained from both papers show that there are surprisingly substantial differences in the outcomes of the experiments if NTBs are modelled with the help of import tariffs or technological change variables, although the two approaches tend to affect the terms of trade in a similar manner. The authors emphasize that the use of the import tariff approach to model NTBs and the corresponding artificial rent-creating and tariff revenue mechanism require a very careful analysis of the resulting welfare effects (FUGAZZA and MAUR, 2008). The authors also conclude that the efficiency modelling of NTBs tends to weigh heavily in the overall large, positive welfare gains. CHANG and HAYAKAWA (2010), PHILIPPIDIS and CARRINGTON (2005), PHILIPPIDIS and SANJUÁN (2006, 2007) and WINCHESTER (2009) obtained the same results using estimated AVEs of NTBs in a CGE model applying the iceberg cost approach. Based on these findings, the iceberg cost approach is utilized for the simulations with the GTAP model.

Experiment Design

In the following GTAP analysis, version 8 of the data base is employed and aggregated into 35 sectors and 20-regions aggregated version. In so doing, major trading partners of the EU and US as well as other countries that are currently involved in FTAs with EU and US are singled out. In the sector aggregation process, the sectors that are predefined in the gravity model approach are matched. Hence, all available food and agricultural sectors are used and the non-food sector is split into 10 manufacturing sectors and nine services sectors. Countries, regions and sectors are highlighted in more detail in Table A1 in the Appendix.

The base year of the GTAP database (version 8) is 2007. The global database therefore already captures many of the changes that took place in the trade policy environment during the last decades. The Multi-Fibre Arrangement (MFA), phased out in 2005, and the expansion of the EU to include Bulgaria and Romania (2007) took place. China is a member of the WTO fulfilling its scheduled obligations.

For the analysis of the TTIP, a baseline projected from the benchmark year 2007 to 2020 is developed. Here, we also consider the ongoing and recently concluded bilateral trade negotiations of the EU and the US as well as the EU enlargement to include Croatia (2013). Besides the change in the political environment, we additionally take the economy-wide changes of the macroeconomic variable into account by updating the factor endowment and population in each country and region to the year 2020. Thereby, the model endogenously generates the value of the technical change parameter necessary to reach the projected growth rates of the prevailing economies.

This baseline is compared to a policy scenario in which we completely abolished tariffs and cut NTBs by 50% between the EU and the US. In so doing, we assume that the EU and US manage to harmonize regulations and standards by 50% until 2020. Additionally, we assume that the change of the political and economic environment will be completely implemented within the given time period. This includes the above-mentioned change in tariffs and NTBs of the EU and the US, but also resulting spill-over effects in third countries.[8]

Simulation Results

Table 2 demonstrates the welfare effects of the TTIP in the EU, the US and third countries. The TTIP leads to substantial gains for the EU and the US of 105 and 156 $ US billion, respectively. In contrast, the TTIP results in negative welfare effects for third countries, even though we assume a positive spill-over effect to take place within the considered time period.

In Table 2, we further decompose the total result. Here, we differentiate between the initializing factor of the welfare effect (tariffs or NTBs) and the corresponding bilateral trade flow (e.g., exports from the US to the EU are reported in column "USA → EU"). Summarizing the main results from Table 2, we conclude:

- As expected, the abolishment of tariffs within the TTIP results in welfare losses for the tariff cutting country, whereas the FTA partner experiences welfare gains. For example, the dismantling of EU tariffs causes a welfare loss of 7.4 $ US billion for the EU, whereas the US experiences a welfare gain of 15.98 $ US billion.
- In general, NTBs are much more important than tariffs for the welfare effects. Table 2 shows that most of the welfare effects caused by tariff cuts are less than 10% of the total welfare effect.

---

8 In accordance with CEPR (2013) we model the spill-over effects also as a reduction in trade costs or a gain in efficiency. A direct spill-over effect is given when third countries improve their market access to the EU and the US. We assume the direct spill-over effect to be 10% of the estimated NTBs between the EU and the US. An indirect spill-over effect captures the improved access of the EU and the US to the market of third countries which is assumed to be equal to 50% of the direct spill-effect. Finally, we also assume a trade promoting effect of the TTIP between third countries of 0.01%.

Impacts of an EU-USA-Free Trade Agreement on Developing Countries   121

Tab. 2: Welfare Changes of the TTIP in the EU, the US and Third Countries (Equivalent Variation, $ US Billion)

	Total	EU				USA				Third Countries (TC)
		Tariffs	NTBs			Tariffs	NTBs			NTBs
			USA→EU	TC→EU	EU→TC		EU→USA	TC→USA	USA→TC	TC→TC
EU	105.18	-7.40	20.24	22.46	6.82	8.03	58.69	-2.77	-0.85	-0.05
USA	156.12	15.98	71.30	-2.23	-0.48	-3.08	50.27	18.84	5.56	-0.03
HIC	-15.99	-2.81	-13.08	4.57	1.56	-0.79	-10.89	4.02	1.13	0.30
China	-10.40	-1.92	-9.00	2.59	0.90	-0.77	-5.07	2.65	0.03	0.19
India	-2.34	-0.75	-3.07	1.37	0.35	-0.06	-0.87	0.68	-0.06	0.07
Brazil	-2.42	-0.35	-1.72	0.44	0.10	-0.09	-0.98	0.23	-0.07	0.02
LDC	-1.56	-0.01	-0.25	-0.05	0.23	-0.23	-1.19	-0.14	0.06	0.01
ROW	-19.23	-1.62	-10.12	2.50	3.28	-1.74	-13.81	1.04	0.98	0.25
World	209.35	1.12	54.30	31.63	12.76	1.27	76.16	24.55	6.77	0.77

Source: Own calculation

- The welfare effects due the adjustment of technical standards and elimination of unnecessary restrictive NTBs is roughly equally important for the EU and the US.
- The spill-over effects caused by the reduction of NTBs with third countries result in positive welfare effects for the NTB reducing country, while the FTA partner face welfare losses. For example, the EU gains 22.46 $ US billion when third countries obtain an improved access to the EU market according to the spill-over effect. The same effect causes a welfare loss for the US which amounts to -2.23 $US.
- Third countries are able to gain mostly from the adoption of a common standard for NTBs when they trade with the EU and the US, but also with each other.

Table 3 reports the changes of the trade balance for selected food and agricultural products. The TTIP leads to an increase of EU imports of most food and agricultural products relative to the corresponding exports. Exceptions to this are given for the dairy sector and for the vegetable oils and fats sector. In contrast to that, the results show an increase of the US trade balance for cereal grains, fruits and vegetable, processed rice, cattle beef and other meat. Table 3 also shows the effects of the TTIP for third countries, which are overall noteworthy.

The upper part of Table 3 shows the total effect of the TTIP on the trade balance in the prevailing country. In line with the previous approach, we decompose this total effect according to the initiating factor (tariffs or NTBs) and report the results in the middle and lower part of Table 3. The results confirm that the elimination of NTBs is much more important than the abolishment of tariffs. A 50% reduction of NTBs leads to a considerable increase of relative imports in most of the processed food sectors in the EU. However, the elimination of NTBs does not cause an increase of the EU trade balance either. We rather observe a reduction in the sector for vegetable oils and fat, cattle meat, dairy, other meat and other food products. In contrast, many of third countries' relative exports are increased due to the spill-over effect of the TTIP, so that their trade balances show an increase.

Tab. 3: Change in the Trade Balance for Selected Food and Agricultural Products ($ US Million)

	EU	USA	HIC	China	India	Brazil	LDC	ROW
Total								
Wheat	-238.0	-17.4	49.9	6.5	15.2	14.6	24.6	167.8
Cereal grain	-413.0	181.0	49.5	32.2	-0.1	14.2	11.5	166.6
Oil seeds	-245.0	-54.1	47.3	210.0	7.8	22.6	9.6	98.5
Fruit and vegetable	-2179.0	999.0	105.9	169.0	168.0	37.3	61.7	386.0
Sugar	-203.0	-56.9	17.8	11.5	-6.3	6.9	40.9	187.0
Vegetable oils and fats	164.0	-458.0	27.7	227.0	93.2	51.3	54.3	-116.4
Processed rice	-121.0	96.2	5.6	5.8	-27.2	4.7	58.6	-8.5
Cattle meat	-806.0	96.6	446.3	14.6	-17.7	46.7	12.1	203.2
Dairy	740.0	-768.0	-178.8	30.5	-12.7	3.5	28.8	135.8
Other meat	-1418.0	448.0	283.1	230.0	2.5	125.0	37.0	291.2
Other food products	-4901.0	-882.0	2121.8	550.0	73.4	163.0	262.7	2043.5
Tariffs								
Wheat	-103.0	-18.5	27.0	1.7	3.7	4.6	7.0	85.0
Cereal grain	-9.4	-88.4	19.1	12.6	0.2	-7.5	3.3	84.3
Oil seeds	38.2	-175.0	22.3	26.1	2.1	39.3	3.2	65.4
Fruit and vegetable	-186.0	-105.0	50.0	41.3	52.0	2.0	18.9	113.1
Sugar	-6.6	-9.6	3.6	1.6	-1.6	-7.3	3.8	16.2
Vegetable oils and fats	95.6	-118.0	3.9	33.2	19.3	-4.7	8.5	-25.2
Processed rice	-64.6	72.8	2.9	0.5	-11.6	0.1	9.1	-4.9
Cattle meat	-478.0	422.0	62.1	2.3	-3.6	-12.9	1.5	9.8
Dairy	557.0	-417.0	-148.0	3.7	-10.1	-0.5	3.5	-14.7
Other meat	-1199.0	1054.0	42.6	53.9	0.5	-6.5	6.2	39.9
Other food products	-2756.0	4103.0	-404.3	-67.5	-33.7	-32.9	-28.1	-911.6
Non-Tariff Barriers								
Wheat	-135.0	1.1	23.0	4.8	11.5	10.0	17.6	82.5
Cereal grain	-404.0	269.0	30.3	19.6	-0.2	21.6	8.2	82.1
Oil seeds	-283.0	121.0	24.9	184.0	5.7	-16.7	6.4	33.1
Fruit and vegetable	-1994.0	1105.0	56.0	128.0	116.0	35.4	42.7	273.2
Sugar	-197.0	-47.3	14.1	9.9	-4.8	14.2	37.0	170.7
Vegetable oils and fats	68.6	-340.0	23.9	194.0	73.9	55.9	45.7	-91.6
Processed rice	-56.5	23.4	2.7	5.3	-15.7	4.6	49.6	-3.6
Cattle meat	-328.0	-325.0	383.6	12.3	-14.2	59.6	10.7	193.4
Dairy	183.0	-350.0	-31.4	26.8	-2.6	4.0	25.3	150.3
Other meat	-219.0	-606.0	239.8	176.0	1.9	132.0	30.9	251.2
Other food products	-2145.0	-4986.0	2527.4	618.0	107.0	196.0	291.3	2956.9

Source: Own calculation

If we compare the effect of the reduction of tariff and NTBs in the TTIP partner countries and third countries, we observe a more evenly distributed effect for the elimination of NTBs world-wide. In contrast, the effect of the abolishment of tariffs between the TTIP partner countries and the corresponding trade diverting effect is not so pronounced in third countries.

Summary

A successful conclusion of the TTIP between the EU and the US will create one of the biggest FTAs world-wide between two already very important trading blocks. If it is considered from a strategic angle, the conclusion of the WTO agreement in Bali in December 2013 was a very important pathfinder for the TTIP. Compliance with the WTO requirements has gained in importance with the Bali agreement. Thus, it seems to be clear that a TTIP will not be possible on the long-term without a reduction of tariffs and NTBs in the food and agricultural sector. On the contrary, the TTIP might be an option to set a first step towards global standards in many sectors.

The analysis in this paper is based on an integrated econometric-CGE analysis, where econometrically estimated NTBs are used in conjunction with the GTAP model. Thereby, we estimate NTBs for a disaggregated food and agricultural sector comprising 16 primary agricultural and processed food sectors. The results show that the elimination of NTBs between the EU and US is of substantially higher importance for TTIP than the abolishment of tariffs. The EU and the US will experience a noteworthy increase of their welfare and GDP, whereas we notice a negative development for these variables in third countries. However, a clear trend in the results for the food and agricultural sectors of the two partners of the TTIP cannot be observed. None of the TTIP countries can consistently increase their exports of food and agricultural products, nor does one show a consistent decrease in imports of these sectors. On the contrary, it seems to be clear that third countries are more likely to gain from the TTIP if they adapt to the standard set by the EU-US FTA and take advantage of the resulting spill-over effect.

List of References

Anderson, J.E. and van Wincoop, E. (2003): Gravity with gravitas: a solution to the border puzzle. American Economic Review, 93, 170-192.

Anderson, J.E. and van Wincoop, E. (2004): Trade costs. Journal of Economic Literature XLII (3), 691-751.

Anderson, J.E. (2011): The Gravity Model. The Annual Review of Economics, 3 (1), 133-160.

Andriamananjara, S., Ferrantino, M. and Tsigas, M. (2003): Alternative approaches in estimating the economic effects of non-rariff measures: Results from newly quantified measures. Working Paper 2003-12-C, U.S. International Trade Commission.

Andriamananjara, S., Dean, J.M., Feinberg, R., Ferrantino, M., Ludema, R. and Tsigas, M. (2004): The effects of non-tariff measures on prices, trade, and welfare: CGE implementation of policy-based price comparisons. Working paper 2004-04-A, U.S. International Trade Commission.

Brockmeier, M. (2003): Ökonomische Auswirkungen der EU-Osterweiterung auf den Agrar- und Ernährungssektor, Simulationen auf der Basis eines Allgemeinen Gleichgewichtsmodells. Habilitationsschrift, Wissenschaftsverlag Vauk Kiel, Agrarökonomische Studien, Band 22.

Brockmeier, M. and Pelikan, J. (2008a): Agricultural market access: A moving target in the WTO negotiations? Food Policy, 33, 250-259.

Brockmeier, M. and Pelikan, J. (2008b): Die WTO-Verhandlungen und die GAP, Agrarwirtschaft. 57 (3/4), 165-178.

Brockmeier, M. and Pelikan, J. (2007): WTO-Verhandlungen: Warum die Einigung so schwierig und dennoch so wichtig ist. Agrarwirtschaft, 56 (3), 145-146.

Centre for Economic Policy Research (CEPR, ed.) (2013): Reducing Transatlantic Barriers to Trade and Investment. An Economic Assessment. Final Report of the Study for the European Commission.

Chang, K. and Hayakawa, K. (2010): Border barriers in agricultural trade and the impact of their elimination: Evidence from East Asia. The Developing Economies, 48 (2), 232-246.

Chen, N. (2004): Intra-national versus international trade in the European Union: Why do national border matter? Journal of International Economics, 63 (1), 93-118.

Chen, N. and Novy, D. (2012): On the measurement of trade costs: direct vs. indirect approaches to quantifying standards and technical regulations. World Trade Review, 11 (3), 401-414.

EU Commission (2013): Trade Statistics, Trade picture files by country, EU28 Agricultural Trade with USA. In: http://ec.europa.eu/agriculture/trade-analysis/statistics/outside-eu/ 2013 /us-factsheet_en.pdf [last access: January 2014].

EU Commission (2014): In focus: Transatlantic Trade and Investment Partnership (TTIP). In: http://ec.europa.eu/trade/policy/in-focus/ttip/ [last access: January 2014].

Feenstra, R. (2002): Border effects and gravity equation: consistent methods for estimation. Scottish Journal of Political Economy, 49 (5), 491-506.

Felbermayr, G., Larch, M., Flach, L., Yalicin, E. and Benz, S. (2013): Dimensionen und Auswirkungen eines Freihandelsabkommens zwischen der EU und den USA. Final Report of the Study for the Bundesministeriums für Wirtschaft und Technologie, ifo Institut.

Francois, J.F. (1999): Economic effects of a new WTO Agreement under the Millennium Round. Report to the European Commission Trade Directorate.

Francois, J.F. (2001): The next WTO Round: North-South stakes in new market access negotiations. Centre for International Economic Studies, Adelaide, ISBN 0-86396 474-5.

Fugazza, M. and Maur, J.-C. (2008): Non-tariff barriers in CGE models: How useful for policy? Journal of Policy Modeling, 30 (3), 475-490.

GTAP (2014): GTAP 6 Data Base Beta Release. In: https://www.gtap.agecon.purdue.edu/ databases/v6beta/v6b_doco.asp [last access: January 2014].

Head, K. and Mayer, T. (2013): Gravity Equations: Workhorse, Toolkit, and Cookbook. In: Gopinath, G., Helpman, E., Rogoff, K. (eds.), Handbook of International Economics, 4.
Hertel, T.W. (ed.) (1997): Global Trade Analysis. Modeling and Applications, New York.
Hertel, T.W., Walmsley, T. and Itakura, K. (2001a): Dynamic effects of the "New Age" Free Trade Agreement between Japan and Singapore. GTAP Working Paper 15.
Hertel, T.W., Walmsley, T.L. and Itakura, K. (2001b): Dynamic Effects of the "New Age" Free Trade Agreement between Japan and Singapore. Journal of Economic Integration, 16 (4), 446-448.
Koester, U. and Schmitz, P.M. (1982): The EC sugar policy and developing countries. European Review of Agricultural Economics, 9, 183-204.
Olper, A. and Raimondi, V. (2008): Agricultural market integration in the OECD: A gravity-border effect approach. Food Policy, 33, 165-175.
Pelikan, J. (2009): Quantitative Analysen zu den WTO-Agrarverhandlungen der Doha-Runde. Dissertation, Justus-Liebig-Universität Giessen.
Philippidis, G. and Carrington, A. (2005): European enlargement and single market accession: A mistreated Issue. Journal of Economic Integration, 20 (3), 543-566.
Philippidis, G. and Sanjuán, A.I. (2006): An examination of Morocco's trade options with the EU. Journal of African Economics, 16 (2), 259-300.
Philippidis, G. and Sanjuán, A.I. (2007): An Analysis of Mercosur's regional trading agreements. World Economy, 30 (3), 504-531.
Philippidis, G., Resona-Ezcaray, H. and Sanjúan-López, A.I. (2013): Capturing zero-trade values in gravity equations of trade: an analysis of protectionism in agro-food sectors. Agricultural Economics, 00, 1-19.
Samuelson, P.A. (1954): The transfer problem and transport costs II: Analysis of effects of trade impediments. The Economic Journal, LXIV, 264-289.
Santos Silva, J.M.C. and Tenreyro, S. (2011): Further simulation evidence on the performance of the Poisson pseudo-maximum likelihood estimator. Economic Letters, 112, 220- 222.
Santos Silva, J.M.C. and Tenreyro, S. (2006): The log of gravity. The Review of Econonmics and Statistics, 88, 641-658.
Schmitz, P.M. and Koester, U. (1981): Der Einfluss der EG-Zuckerpolitik auf die Entwicklungsländer. Diskussionsbeitrag Nr. 42, Institut für Agrarpolitik und Marktlehre, Christian-Albrechts-Universität Kiel.
Schmitz, P.M. (1982): Exporterlösstabilisierung durch AKP-Präferenzabkommen. Diskussionsbeitrag Nr. 47, Institut für Agrarpolitik und Marktlehre, Christian-Albrechts-Universität Kiel.
Schmitz, P.M. (1984): Handelsbeschränkungen und Instabilität auf Weltagrarmärkten. Weltwirtschaftliche Studien, Heft 21, Institut für Europäische Wirtschaftspolitik, Universität Hamburg.
Winchester, N. (2009): Is there a dirty little secret? Non-tariff barriers and the gains from trade. Journal of Policy Modeling, 31, 819-834.
WTO (2013): http://rtais.wto.org/UI/PublicAllRTAList.aspx [last access: December 2013].

Appendix

Tab. A1: Sector and Region Aggregation

Regions	Sectors
1 European Union Austria, Belgium, Denmark, Finland, France, Germany, Ireland, United Kingdom, Greece, Italy, Luxembourg, Netherlands, Portugal, Spain, Sweden, Czech Republic, Hungary, Malta, Poland, Slovakia, Slovenia, Estonia, Latvia, Lithuania, Cyprus, Romania, Bulgaria, Croatia	1 Paddy rice
2 USA	2 Wheat
3 Canada	3 Cereal grains
4 Japan	4 Vegetables and fruits
5 Korea	5 Oil seeds
6 China	6 Sugar cane, sugar beet
7 India	7 Plant-based fibres
8 Brazil	8 Crops
9 Mexico	9 Cattle
10 Turkey	10 Other animal products
11 Bangladesh	11 Raw milk
12 Mozambique	12 Sugar
13 Other high income countries Australia, New Zealand, Hong Kong, Taiwan, Singapore, Switzerland, Norway, Rest of EFTA	13 Processed rice
14 Central America Costa Rica, Guatemala, Honduras, Nicaragua, Panama, El Salvador, Rest of Central America, Caribbean	14 Dairy
15 East Europe Albania, Belarus, Ukraine, Rest of Eastern Europe, Rest of Europe	15 Cattle meat
16 North Africa Israel, Rest of Western Asia, Egypt, Morocco, Tunisia, Rest of North Africa	16 Other meat
17 ASEAN Indonesia, Malaysia, Philippines, Thailand, Viet Nam,	17 Vegetable oils and fats
18 Mercosur Argentina, Paraguay, Uruguay	18 Other food products
19 Low income countries Cambodia, Lao People's Democratic Republic, Rest of Southeast Asia, Nepal, Rest of South Asia, Benin, Burkina Faso, Guinea, Senegal, Togo, Rest of Western Africa, Central Africa, South Central Africa, Ethiopia, Madagascar, Malawi, Rwanda, Tanzania, Uganda, Rest of Eastern Africa	19 Beverages and tobacco
20 Rest of the World Rest of Oceania, Mongolia, Rest of East Asia, Pakistan, Sri Lanka, Rest of North America, Bolivia, Chile, Colombia, Ecuador, Peru, Venezuela, Rest of South America, Russian Federation, Kazakhstan, Kyrgyztan, Rest of Former Soviet Union, Armenia, Azerbaijan, Georgia, Bahrain, Iran Islamic Republic of, Kuwait, Oman, Qatar, Saudi Arabia, United Arab Emirates, Cameroon, Cote d'Ivoire, Ghana, Nigeria, Kenya, Mauritius, Zambia, Zimbabwe, Botswana, Namibia, South Africa, Rest of South African Customs, Rest of the World	20 Other primary sectors Wool and silk-worm cocoons, Forestry, Fishing, Minerals nec
	21 Other primary energy Coal, Oil, Gas
	22 Chemicals
	23 Electrical machinery
	24 Motor vehicles
	25 Other transport equipment
	26 Other machinery
	27 Metals and metal products Ferrous metals, Metals nec, Metal products
	28 Wood and paper products Wood products, Paper products and publishing
	29 Other manufactures Textiles, Leather products, Petroleum and coal products, Mineral products nec, Manufactures nec
	30 Water transport
	31 Air transport
	32 Finance
	33 Insurance
	34 Business services
	35 Communications
	36 Construction
	37 Personal services
	38 Other services Electricity, Gas manufacture and distribution, Water, Trade, Transport nec, PubAdmin/Defence/Health/Education, Dwellings

Tab. A2: Estimation Results Pooled across Sectors

	Agriculture	Manufacturing	Services	All
ln(Distance)	-0.622***	-0.521***	-0.247***	-0.480***
	(0.0281)	(0.0263)	(0.0244)	(0.0223)
Landlocked Status	-0.734***	-0.230	-0.645**	-0.355**
	(0.230)	(0.155)	(0.311)	(0.157)
Island	0.196	0.385*	-0.198	0.301*
	(0.214)	(0.209)	(0.161)	(0.181)
Common Land Border	0.513***	0.353***	0.430***	0.356***
	(0.0759)	(0.0746)	(0.0824)	(0.0679)
Common Language	0.409***	0.244***	0.188***	0.265***
	(0.0846)	(0.0772)	(0.0631)	(0.0681)
Common Colonizer	0.978***	0.299*	0.260***	0.348**
	(0.107)	(0.174)	(0.0805)	(0.144)
$\ln(1+AVE_{Tariff})$	-0.234	-2.247***		-0.936***
	(0.186)	(0.760)		(0.337)
EU	1.003***	0.463***	0.210**	0.438***
	(0.108)	(0.0944)	(0.0842)	(0.0799)
NAFTA	0.437**	1.028***	0.385***	0.908***
	(0.222)	(0.135)	(0.129)	(0.118)
AUSNZL	0.959***	1.718***	-0.292*	1.343***
	(0.319)	(0.235)	(0.169)	(0.200)
Observations	206112	128820	115938	450870
EU-specific Effect %	172.60	58.82	23.40	54.88

Notes: Importer, exporter and sector fixed effects not reported. Robust standard errors clustered at the country pair level in parentheses. Constant terms are included but not reported. *, **, and *** indicate significance at 10, 5 and 1% levels, respectively.

Source: Own calculation

Most of the coefficients in Table A2 have the expected signs and are in line with results from the literature. As expected, the elasticity of trade with respect to distance is negative. The strongest effect is observed in the food and agricultural sector. Hence, if the distance between two countries increases by 1%, bilateral trade in food and agriculture decreases by 0.62%. If one of the countries is landlocked, it influences trade negatively. The effect of being landlocked is highly significant except in the manufacturing sector. In contrast, the effect of island status is positive and only significant in the manufacturing and all sector regressions. As expected, contiguity and cultural adjacency increase trade significantly in all sector regressions. Also, tariffs have a negative effect on trade levels. The negative effect is highly significant in the manufacturing sector and in the all sector regression. If tariffs increase by 1%, trade in manufacturing decreases by 2.25%. Finally, the most important result for our analysis is the parameter of the EU-specific dummy variable. Consistent with expectations, intensive trade relations within mutual agreements increase trade significantly. Hence, if both countries are in the EU and implicitly have gone through the process of mutual recognition or harmonization of regulations and standards, trade in agro-food increases by 172.60%, in manufacturing by 58.82% and in services by 23.40%. Considering all sectors, the EU-specific effect amounts to 54.88%.

129

EU-South Korea Free Trade Agreement and Its Impacts on Agriculture in Consideration of a Different Level of Regional Aggregation: A Computable General Equilibrium Approach

Jong-Hwan KO

Abstract

In July 2011 the EU-Korea free trade agreement (FTA) entered into force. This paper aims at assessing its economic effects on the economies of Korea and Germany, in particular, on their agricultural sectors, using a multi-region, multi-sector CGE model taking into consideration different levels of regional aggregation of the EU. The impacts on Korea's economy vary with regard to the level of regional disaggregation of the EU. The analyses accounts for all EU member states but Croatia. The latter is part of the regional grouping called Rest of the World. Impacts of implementing the EU-Korea FTA are simulated using several scenarios which differ as regards the aggregation of the EU in the model. The two extreme cases are scenario EU1 (the EU is represented as a single region) and scenario EU27 (all 27 member states are represented individually). In between these two extremes other scenarios are set up which reflect additional levels of EU disaggregation in the model. The simulations results of each of the various scenarios are compared to a corresponding baseline scenario reflecting the same level of EU disaggregation. Not to complicate the discussion too much the results of only the two extreme scenarios are briefly contrasted. For this comparison the relative deviations of each of the two scenarios to its corresponding baseline are taken. The higher the level of regional disaggregation of the EU the stronger is the increase in Korea's GDP and welfare due to the implementation of the EU-Korea FTA. Simulation results for scenario EU27 (highest disaggregation) show an additional increase of 44.44% and 38.16% in the two indicators, respectively, compared to scenario EU1 (lowest disaggregation). Similar outcomes are found for Korea's export and import volumes; i.e. the higher the EU is disaggregated the more increase these trade volumes are simulated by the model. The opposite result is observed for domestic production of most of Korea's agricultural sectors. These sectors shrink somewhat and at various degrees if regional disaggregation of the EU is increased in the model. However, the results for Germany deviate from those of Korea. The difference in the level of regional aggregation of the EU does not lead to varying impacts on the real GDP and welfare of Germany as well as on this country's total export and import volumes. The findings of this paper indicate that the most disaggregated data of 27 member states of the EU should be used for an analysis of the impacts of the EU-Korea FTA, in particular, on the Korean economy and that all the previous studies on the EU-Korea FTA that used the EU as a single aggregated entity may have underestimated the potential effects of the EU-Korea FTA, in particular, on the Korean economy.

Acknowledgements

The author would like to thank Prof Dr Thomas HECKELEI and Dr Wolfgang BRITZ for helpful discussions on some of the issues discussed in this paper, and Prof Dr Klaus FROHBERG for constructive comments. Any remaining errors are the author's.

Introduction

After eight rounds of negotiations of the EU-Korea free trade agreement (FTA) since the Republic of Korea (henceforth referred to as Korea) and the European Commission declared its launching on 6 May 2007, the free trade agreement signed by both parties on 6 October 2010 entered into force on 1 July 2011. The EU-Korea FTA is the 6th FTA that Korea has established so far[1], since the Korean government drew up the FTA roadmap in 2003 in an effort to enhance international competitiveness for the country's trade-dependent economy.

As of December 2013, the EU has a preferential trade agreement (PTA) or an economic partnership agreement in place with 49 countries of the world[2]. The number of the countries with which the EU is currently negotiating a preferential trade agreement or a PTA pending official conclusion is 84. The number of countries with which the EU is considering opening negotiations about preferential agreement is 6 and the EU is negotiating a stand-alone investment agreement with China. The EU-Korea FTA is the first trade agreement that the EU concluded with any country in East Asia.

In 2012, bilateral trade between the EU and Korea reached €75.6 billion (The European Commission, 2013a). The EU remains Korea's 2nd largest trading partner after China, with the position of the 2nd largest destination of Korea's exports, while it has become the 4th largest source of Korea's imports, next to China, Japan and the United States. Korea is the 10th largest trading partner of the EU. The EU is the biggest investor to Korea. Therefore, the EU-Korea FTA is expected to affect the economies of the EU and Korea.

It is of great importance to estimate the potential economic effects of the EU-Korea FTA for both economies. Some empirical studies on its effects have been conducted: COPENHAGEN ECONOMICS & FRANCOIS (2007), DECREUX, MILNER AND PÉRIDY (2010), KIM et al. (2005), KO (2006, 2007, 2014A, 2014B), and KO and LEE (2008). All of these studies used a computable general equilibrium (CGE) model. There are three types of CGE models that can be used for an impact analysis of an FTA[3]: a static CGE model, a recursively dynamic

---

1   As of January 2014, Korea has 9 FTAs in effect with 48 countries such as Chile, Singapore, EFTA (European Free Trade Association), ASEAN (Association of Southeast Asian Nations), India, EU (European Union), Peru, the United States and Turkey, as seen in Table A1 of the appendix of this chapter.
2   In addition, the EU has customs union with 4 countries suche as Andorra, Monaco, San Marino and Turkey. For detailed information, see the European Commission (2013b).
3   A static CGE model is one in which time cannot be explicitly considered. A recursively dynamic CGE model is one in which a static equilibrium for each period is calculated. From an empirical perspective, a recursively dynamic CGE model for an impact analysis of an FTA is one in which the timing of policy implementation is explicitly considered. A

CGE model, and a forward-looking dynamic CGE model (KO, 1993). COPENHAGEN ECONOMICS & FRANCOIS (2007), KIM et al. (2005) and KO (2014b) used a static CGE model for an impact analysis of the EU-Korea FTA and KO (2006, 2007, 2014a) and KO and LEE (2008) used a recursively dynamic CGE model for it.

In such studies, a decision on the aggregation level with regard to regions and sectors is needed. Quite often, that decision is made a priori before the model runs. There is ample evidence that sector aggregation matters (GRANT, HERTEL and RUTHERFORD, 2006; LENZEN, 2011; PELIKAN and BROCKMEIER, 2008; ZHANG et al, 2013). However, there is limited evidence about how the regional aggregation level affects the results of CGE applications. We take the EU-Korea FTA as an example to analyze how the regional aggregation level will affect the results, because the EU consists of 28 member states.

Against this backdrop, the objective of this study is to conduct a quantitative assessment of the economic effects of the EU-Korea FTA on agricultural sectors using a static[4] multi-region, multi-sector CGE model taking into consideration different levels of regional aggregation of the EU. The EU consisting of 28 member states is a good example for such an experiment. The European Community (EC)[5] started with the 6 founding members, Belgium, France, Italy, Luxembourg, the Netherlands and West Germany in the form of the European Coal and Steel Community (ECSC) established in 1951 based on the Treaty of Paris. The European Community has enlarged over time from 6 member states in 1951 to 28 in 2013. In more detail, 6 member states in 1951, 9^6 in 1973, 10^7 in 1981, 12^8 in

---

forward-looking dynamic CGE model is one in which perfect foresight is assumed. However, the assumption of a forward-looking expectation for an inter-temporally dynamic CGE model is not so realistic for policy implementation.

4   For simplicity's sake, a static CGE model is used for this study, although it is more desirable to use a recursively dynamic CGE model for an impact analysis of the EU-Korea FTA, which makes it possible to take into account a different timing of concessions of tariff reductions agreed on in the EU-Korea FTA and capital accumulation over time. In order to use a recursively dynamic CGE model, not only forecasts of key macroeconomic variables such as the growth rates of GDP, skilled labor, unskilled labor, and population of each region, but also policy changes other than the EU-Korea FTA that have already taken place or are anticipated to take place in the future are needed for baseline scenarios. However, it is too much time consuming to obtain such baseline scenarios, for instance, for scenario 8 with 44 regions, as seen in Table 2. Therefore, a static CGE model is used for this study.
5   The abbreviations EC and EU are used interchangeably in this paper. The term EU is applied since November 1993 when the Treaty of Maastricht became effective.
6   Denmark, Ireland and the United Kingdom joined the EC in 1973.
7   Greece became a member of the EC in 1981.
8   Portugal and Spain joined the EC in 1986.

1986, 15[9] in 1995, 25[10] in 2004, 27[11] in 2007 and 28[12] in 2013. Such a step-by-step increase in its membership is considered in policy scenarios for the EU-Korea FTA to measure to what extent a different level of regional aggregation will affect its simulation results.

The remainder of this study is organized as follows. The first section describes the CGE model and data used in this study. The second section provides a brief description of trade relations between the EU and Korea. The third section examines baseline and policy scenarios of the EU-Korea FTA, and the last section discusses their simulation results before concludes with some remarks.

The CGE Model and Data

The CGE Model

The CGE model used in this study is an extended version of the static[13] GTAP model (HERTEL, 1997), which incorporates the interaction between trade liberalization and capital accumulation based on classical growth theory (BALDWIN, 1989 and 1992).

According to this theory, a medium-run growth induces additional savings and investment, which yields more output. In other words, the initial increase in income, as a result of trade liberalization, is to make savings and investment grow. These induced savings and investment, thus larger capital stock, lead to larger production capacity and cause a further increase in income[14].

As seen in Figure 1, each regional[15] economy includes as economic agents producers, consumers, the government and a regional household. Producers are included in the model as a single firm for each sector and consumers as a private household. The regional household is unique in the GTAP model. It is assumed that the regional household collects all factor incomes and taxes generated in the

---

9   In 1995, the EU enlarged by granting Austria, Finland and Sweden membership.
10 8 Central and Eastern European countries (CEECs) such as Czech Republic, Estonia, Hungary, Latvia, Lithuania, Poland, Slovakia and Slovenia as well as Cyprus and Malta joined the EU in 2004.
11 Bulgaria and Romania became members of the EU in 2007.
12 Croatia joined the EU in 2013.
13 See KO (2014a) for a dynamic CGE model used for an impact analysis of the EU-Korea FTA which makes it possible to explicitly take into account capital accumulation over time and the different timing of policy implementations of the EU-Korea FTA.
14 Nonetheless, the model is not a recursively dynamic, but a static one, because capital accumulation over time is not explicitly considered.
15 A region represents a single country or an aggregate of several countries.

region and spends these incomes on three components of final demand, i.e. private consumption, government expenditure and savings, according to a Cobb-Douglas utility function[16].

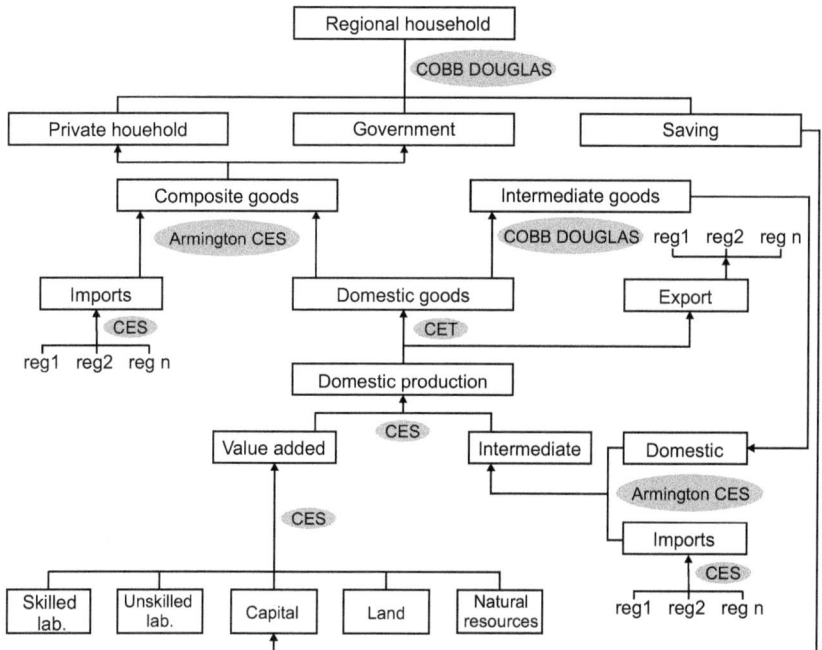

Fig. 1:   Structure of the CGE Model
Source: Own presentation based on the GTAP model

The CGE model has solid micro-foundations that are theoretically transparent. Consumers maximize their utility subject to budget constraint. Producers minimize production costs under consideration of production technology. Production structures are represented by nested production functions consisting of Cobb-Douglas and Constant Elasticity of Substitution (CES) functions, as seen in Figure 1. Perfect competition and constant returns to scale are assumed in production.

---

16 This assumption has some advantages as well as some disadvantages. The most significant drawback is the failure to link government expenditures to tax revenues. The greatest advantage of the assumption is the unambiguous indicator of welfare offered by the regional utility function. See HERTEL (1997) for more details.

For manufacturing a given level of output, a firm is assumed to minimize costs by purchasing primary factors of production, and intermediate inputs and by supplying output to both domestic and export markets via a Constant Elasticity of Transformation (CET) function in response to prices in commodity and factor markets. The model distinguishes as factors of production unskilled labor, skilled labor, capital, land and natural resources. Labor and capital are employed by all sectors, but land is used only in agricultural sectors and natural resources are utilized in mining sectors and forestry. It is assumed that intermediate inputs and capital are traded between regions, while labor, land and natural resources are not traded between regions.

Each of the goods demanded is a composite of both domestically manufactured and imported products. As origin of imports all regions considered in the model are included. The government purchases domestic and imported goods and services, based on a Cobb-Douglas aggregation function. Private household expenditure is explained by the Constant Difference of Elasticities (CDE) functional form[17], which lies between the non-homothetic CES on the one hand, and the fully flexible functional forms on the other hand[18]. Investment is driven by savings and does not affect productive capacity but affects total activity as a component of final demand.

Imperfect substitution in goods and services between home and abroad, and imperfect substitution among different origins of economies are modelled employing the ARMINGTON approach (ARMINGTON, 1969), which is often used for depicting intra-industry trade, i.e. two-way trade in the same product category, but originating from different nations. Since traded and non-tradable goods are assumed to be distinct and imperfect substitutes by sector, changes in relative market prices of other regions are only partially transmitted to domestic markets. Thus, the model incorporates a realistic degree of insulation of domestic commodity markets from those prevailing in markets of the other regions.

The model solves for commodity and factor prices that equate demand and supply in all commodity and factor markets. The model also solves for world prices, equating demand and supply of sector exports and imports across the world economy.

The model is of comparative statics in nature given the pattern of world output and trade at one moment of time, and it generates what the pattern of output and trade would be after the world economies adjusted to trade liberalization caused

---

17 The CDE functional form was first proposed by Hanoch (1975).
18 For an exhaustive treatment of the calibration and use of the CED functional form in CGE models, see HERTEL et al. (1991).

EU-South Korea Free Trade Agreement 135

by various policy scenarios according to differences in the specification of the EU-Korea FTA.

Thus, the effects of the EU-Korea FTA to be quantified in this study are static ones. The static effects stem from changes in the allocative efficiency of partners of the EU-Korea FTA made possible by increased specialization in accordance with the law of comparative advantage, due to the liberalized market of the participating nations, with their productive capacity as given.

Data

The GTAP database prelease candidate 8.1 which was released in February 2013 is used for this study. This database rests on information of the year 2007 and includes 134 regions and 57 sectors. As this paper focuses on agriculture, 57 sectors (industries) are aggregated into 25 ones including 14 primary sectors, 8 processed food sectors, one aggregated mining sector, one aggregated manufacturing sector and one aggregated services sector. They are depicted as numbers 1 to 14, 15 to 22, 23, 24 and 25, respectively in Table 1.

For the purpose of the current research, the 134 regions are aggregated in 8 different ways. The distinctions of aggregation accrue only to the EU, as shown in Table 2. For example, EU1, shown in the first column of Table 2, depicts the 27 member states of the EU[19] as one region. EU7, indicated in the second column, means that 27 member states of the EU are aggregated into 7 regions: EC6[20], DIU[21], G[22], EP[23], AFS[24], CEEC10[25] and BR[26]. EU27 in the last column of Table 2 means that the EU is represented by 27 member states, in other words, that the EU is disaggregated into 27 countries in policy scenario 8.

---

19 Croatia that joined the EU in 2013 is not considered as a single entity, but merged into the ROW.
20 EC6 is a region consisting of the 6 founding members of the EC, Belgium, France, Germany, Italy, Luxembourg and the Netherlands.
21 DIU stands for a region comprising Denmark, Ireland and the United Kingdom which joined the EC in 1973.
22 G is Greece joining the EC in 1981.
23 EP represents a region consisting of Spain and Portugal which entered the EC in 1986.
24 AFS stands for a region made up of Austria, Finland and Sweden which joined the EU in 1995.
25 CEEC10 is a region consisting of Czech Republic, Estonia, Hungary, Latvia, Lithuania, Poland, Slovakia, Slovenia, Cyprus and Malta which joined the EU in 2004.
26 BR is a region consisting of Bulgaria and Romania which joined the EU in 2007.

Tab. 1: Sector classification

Number, abbreviation	Description	Number, abbreviation	Description
1, pdr	Paddy rice	14, fsh	Fish
2, wht	Wheat	15, cmt	Meat: cattle, sheep, goats, horse
3, gro	Cereal grains nec	16, omt	Meat products nec
4, v_f	Vegetables, fruit, nuts	17, vol	Vegetable oils and fats
5, osd	Oil seeds	18, mil	Dairy products
6, c_b	Sugar cane, sugar beet	19, pcr	Processed rice
7, pfb	Plant-based fibers	20, sgr	Sugar
8, ocr	Crops nec	21, ofd	Food products nec
9, ctl	Cattle, sheep, goats, horses	22, b_t	Beverages and tobacco products
10, oap	Animal products nec	23, MNG	Mining
11, rmk	Raw milk	24, MNF	Manufacturing
12, wol	Wool, silk-worm cocoons	25, SRV	Services
13, frs	Forestry		

Note: nec stands for not elsewhere classified

Source: GTAP DB prelease candidate 8.1 (February 2013)

Each of the 8 different levels of regional aggregation shown in Table 2 also includes 16 countries/regions with which Korea has established a free trade agreement, is negotiating an FTA or is considering of negotiating an FTA. This makes it possible to take into account besides trade creating effects of the EU-Korea FTA on the Korean economy also its trade diverting impacts: Chile (CHL), EFTA (European Free Trade Association), Association of Southeast Asian Nations (ASEAN), India (IND), Peru (PER), the United States (USA), Turkey (TUR), China (CHN), Japan (JPN), Australia (AUS), New Zealand (NZL), Canada (CAN), Colombia (COL), Gulf Cooperation Council (GCC) and Mexico (MEX). Table A1, shown in the appendix, provides information about the status of the various FTAs that Korea already concluded, is still in the negotiation phase or considers pursuing.

Table 3 shows applied ad valorem tariff rates of Korea and the EU consisting of 27 member states on imports by sector from each other, which will be used for one of the policy scenarios[27], i.e. policy scenario EU1 of the EU-Korea FTA.

---

[27] Bilateral tariff rates between Korea and each of the 27 member states of the EU to be used for Policy scenario 8 of the EU-Korea FTA are shown in Tables A2 and A3 in the appendix.

EU-South Korea Free Trade Agreement 137

Tab. 2: 8 different levels of regional aggregation of the EU27

EU1 for Scenario 1	EU7 for Scenario 2	EU12 for Scenario 3	EU14 for Scenario 4	EU15 for Scenario 5	EU17 for Scenario 6	EU26 for Scenario 7	EU27 for Scenario 8
1 KOR	1 KOR	1 KOR	1 KOR	1 KOR	1 KOR	1 KOR	1 KOR
2 EU27	2 EC6	2 BEL	2 BEL	2 BEL	2 BEL	2 BEL	2 BEL
	3 DIU	3 FRA	3 FRA	3 FRA	3 FRA	3 FRA	3 FRA
	4 G	4 DEU	4 DEU	4 DEU	4 DEU	4 DEU	4 DEU
	5 EP	5 ITA	5 ITA	5 ITA	5 ITA	5 ITA	5 ITA
	6 AFS	6 LUX	6 LUX	6 LUX	6 LUX	6 LUX	6 LUX
	7 CEEC10	7 NLD	7 NLD	7 NLD	7 NLD	7 NLD	7 NLD
	8 BR	8 DIU	8 DNK	8 DNK	8 DNK	8 DNK	8 DNK
		9 G	9 IRL	9 IRL	9 IRL	9 IRL	9 IRL
		10 EP	10 GBR	10 GBR	10 GBR	10 GBR	10 GBR
		11 AFS	11 G	11 G	11 G	11 G	11 G
		12 CEEC10	12 EP	12 PRT	12 PRT	12 PRT	12 PRT
		13 BR	13 AFS	13 ESP	13 ESP	13 ESP	13 ESP
			14 CEEC10	14 AFS	14 AUT	14 AUT	14 AUT
			15 BR	15 CEEC10	15 FIN	15 FIN	15 FIN
				16 BR	16 SWE	16 SWE	16 SWE
					17 CEEC10	17 CYP	17 CYP
					18 BR	18 CZE	18 CZE
						19 EST	19 EST
						20 HUN	20 HUN
						21 LVA	21 LVA
						22 LTU	22 LTU
						23 MLT	23 MLT
						24 POL	24 POL
						25 SVK	25 SVK
						26 SVN	26 SVN
						27 BR	27 BGR
							28 ROU
3 CHL	9 CHL	14 CHL	16 CHL	17 CHL	19 CHL	28 CHL	29 CHL
4 EFTA	10 EFTA	15 EFTA	17 EFTA	18 EFTA	20 EFTA	29 EFTA	30 EFTA
5 ASEAN	11 ASEAN	16 ASEAN	18 ASEAN	19 ASEAN	21 ASEAN	30 ASEAN	31 ASEAN
6 IND	12 IND	17 IND	19 IND	20 IND	22 IND	31 IND	32 IND
7 PER	13 PER	18 PER	20 PER	21 PER	23 PER	32 PER	33 PER
8 USA	14 USA	19 USA	21 USA	22 USA	24 USA	33 USA	34 USA
9 TUR	15 TUR	20 TUR	22 TUR	23 TUR	25 TUR	34 TUR	35 TUR
10 CHN	16 CHN	21 CHN	23 CHN	24 CHN	26 CHN	35 CHN	36 CHN
11 JPN	17 JPN	22 JPN	24 JPN	25 JPN	27 JPN	36 JPN	37 JPN
12 AUS	18 AUS	23 AUS	25 AUS	26 AUS	28 AUS	37 AUS	38 AUS
13 NZL	19 NZL	24 NZL	26 NZL	27 NZL	29 NZL	38 NZL	39 NZL
14 CAN	20 CAN	25 CAN	27 CAN	28 CAN	30 CAN	39 CAN	40 CAN
15 COL	21 COL	26 COL	28 COL	29 COL	31 COL	40 COL	41 COL
16 GCC	22 GCC	27 GCC	29 GCC	30 GCC	32 GCC	41 GCC	42 GCC
17 MEX	23 MEX	28 MEX	30 MEX	31 MEX	33 MEX	42 MEX	43 MEX
18 ROW	24 ROW	29 ROW	31 ROW	32 ROW	34 ROW	43 ROW	44 ROW

Note: nec stands for not elsewhere classified and the number behind 'EU' refers to the number of regions into which the member states of the EU27 are aggregated

Source: GTAP DB prelease candidate 8.1 (February 2013)

Tab. 3: Ad valorem tariff rates of Korea and the EU27 applied on imports from each other (%)

Sector	Korea	EU27
1 Paddy rice	0.00*	87.66[28]
2 Wheat	0.00**	0.00
3 Cereal grains nec	74.66	10.57
4 Vegetables, fruit, nuts	54.61	5.06
5 Oil seeds	22.65	0.00**
6 Sugar cane, sugar beet	0.00**	0.00**
7 Plant-based fibers	0.04	0.00**
8 Crops nec	14.31	8.60
9 Cattle, sheep, goats, horses	11.25	0.00**
10 Animal products nec	6.61	3.72
11 Raw milk	0.00**	0.00**
12 Wool, silk-worm cocoons	5.99	0.00**
13 Forestry	0.33	0.03
14 Fishing	9.20	1.73
15 Meat: cattle, sheep, goats, horse	17.25	0.16
16 Meat products nec	24.61	15.10
17 Vegetable oils and fats	11.14	9.92
18 Dairy products	39.29	9.77
19 Processed rice	0.00**	4.69
20 Sugar	17.31	1.26
21 Food products nec	73.72	9.93
22 Beverages and tobacco products	19.93	14.58
23 Mining	2.24	0.01
24 Manufacturing sectors	5.59	3.45
25 Services	0.00	0.00

*: There are no imports
**: The value of imports at market prices by source (VIMS) is equal to the value of imports at world prices by source (VIWS)

Source: Own calculation from GTAP database prelease candidate 8.1 (February 2013)

---

28 The tariff rate of 87.66% is Austria's one on imported paddy rice from Korea, as can be seen in Table A3 in the appendix.

Ad valorem tariff rates on primary and manufactured commodities are calculated from GTAP data base prelease candidate 8.1.

Tariff rates of zero imply either that there are no imports of the corresponding product in 2007 or that the value of imports of the corresponding product evaluated at market prices is the same as the value of imports of the product evaluated at world prices[29]. For instance, Korea's tariff rate on paddy rice imported from the EU is zero. This implies that there are no imports of paddy rice by Korea from the EU, as seen in Table 3. There are also no tariffs levied on imported services[30] by Korea and by the EU27, because it is not possible to measure tariffs on services.

Trade Relations between the EU and Korea

Table 4 show the matrix of values of imports[31] at costs, insurance and freight (CIF) prices in 2007 by and from each of the 18 countries/regions aggregated to for scenario 1. As seen from this table, of its total imports of US$407.9 billion in 2007, Korea imported goods and services worth $46.8 billion which equal 11.5 percent of its total imports, from the EU27, while the latter imported goods and services worth $67.5 billion from Korea. These are 1.1 percent of its total imports of $6,011 billion. The EU27 is the 4th largest source of Korean imports, after China, Japan and the United States.

Table 5 shows values of bilateral trade by sector between Korea and the EU in 2007. Korea's total exports[32] to and total imports[33] from the EU27 amounted to $65.4 billion and $49.2 billion, respectively, and Korea ran a trade surplus of $16.2 billion against the EU27. It had trade deficits of $2.3 billion in agricultural and processed food sectors including mining and of $2.3 billion in services, while its manufacturing sectors achieved a trade surplus of $20.8 billion. This implies

---

29 Applied tariff rates used for policy scenarios in this paper are calculated by (VIMS/VIWS -1)*100, where VIMS is the value of imports at market price by source and VIWS is the value of imports at world prices by source. VIMS/VIWS is referred to as the power of ad valorem tariff rates. If VIMS equals VIWS, then the power equals one and the tariff rate equals zero.
30 In order to measure the effects of trade liberalization in services, tariff equivalents of services are needed. For the sake of simplicity, trade liberalization in services sectors is not considered in this paper. For the impact analysis of the EU-Korea FTA including trade liberalization in primary and manufacturing sectors as well as in services, see KO (2014a).
31 A matrix of export values at free on board (FOB) prices is not presented in this paper due to limited space.
32 Korea's exports are evaluated at FOB prices.
33 Korea's imports are evaluated at cost, insurance and freight (CIF) prices.

that Korea has a comparative advantage in manufacturing sectors with respect to the EU, whereas it has a comparative disadvantage in agriculture and services. Korea ran trade deficits, in particular, in the following sectors: 16 Meat products nec ($552 million), 22 Beverages and tobacco products ($485 million), 21 Food products nec ($763 million) and 18 Dairy products ($197 million), as can be seen in Table 5.

Tab. 4: *Matrix of Import values at CIF prices for the year 2007 (billion US$)*

	1 KOR	2 EU27	3 CHL	4 EFTA	5 ASEAN	6 IND	7 PER	8 USA	9 TUR
1 KOR	0.0	67.5	2.5	2.4	36.2	6.5	0.6	55.9	5.1
2 EU27	46.8	3653.5	9.8	220.4	109.0	59.1	3.5	465.1	73.6
3 CHL	5.7	18.3	0.0	0.5	2.1	4.7	1.0	10.3	0.5
4 EFTA	4.5	249.9	0.6	2.6	8.9	8.3	0.2	37.4	6.2
5 ASEAN	35.0	151.1	1.1	7.2	160.3	28.8	0.4	126.1	4.4
6 IND	5.2	65.9	0.4	2.7	14.0	0.0	0.3	40.3	2.3
7 PER	1.4	7.0	1.6	2.5	0.3	0.4	0.0	5.3	0.1
8 USA	56.0	350.9	9.8	24.2	67.9	26.9	4.4	0.0	9.2
9 TUR	0.5	73.5	0.1	2.2	1.5	0.5	0.0	7.6	0.0
10 CHN	70.5	338.2	5.4	11.9	97.8	28.9	1.8	340.9	12.9
11 JPN	61.8	121.1	1.9	4.6	82.2	8.5	0.6	155.6	3.5
12 AUS	12.7	25.6	0.3	0.6	17.9	10.5	0.1	13.1	0.6
13 NZL	1.4	7.9	0.1	0.4	3.1	0.4	0.0	4.2	0.1
14 CAN	3.9	43.2	0.7	5.8	5.9	2.8	0.4	316.5	0.9
15 COL	0.2	6.2	0.5	0.7	0.2	0.1	0.9	10.8	0.2
16 GCC	57.0	72.9	0.1	2.4	56.2	46.8	0.1	57.5	4.9
17 MEX	1.1	20.3	1.5	0.8	1.6	1.5	0.8	218.6	0.3
18 ROW	44.2	738.1	14.2	27.2	69.8	55.1	6.1	360.8	45.4
Total	407.9	6011.0	50.6	319.0	735.1	290.0	21.0	2225.8	170.0

Source: Own calculation based on GTAP DB prelease candidate 8.1 (Feb. 2013)

Tab. 4: *(continued): Import matrix at CIF prices in 2007 (billion US$)*

	10 CHN	11 JPN	12 AUS	13 NZL	14 CAN	15 COL	16 GCC	17 MEX	18 ROW
1 KOR	114.6	30.5	5.0	0.9	4.5	1.2	12.9	8.7	65.9
2 EU27	156.4	92.8	43.6	7.7	51.2	6.1	114.7	34.0	633.3
3 CHL	13.7	12.5	0.4	0.0	1.4	0.7	0.5	2.7	11.4
4 EFTA	9.8	9.5	2.3	0.4	6.4	0.4	7.7	1.6	29.8
5 ASEAN	135.0	95.5	29.7	3.6	10.2	0.8	18.5	5.7	79.1
6 IND	19.5	5.8	1.9	0.4	3.4	0.5	20.9	1.1	46.2
7 PER	6.0	3.7	0.2	0.0	2.1	0.6	0.1	0.3	4.0
8 USA	94.4	95.2	25.5	3.6	244.4	8.7	46.0	135.1	209.9
9 TUR	2.0	0.8	0.5	0.2	0.7	0.1	6.3	0.4	33.5
10 CHN	75.7	133.2	25.9	3.3	32.8	3.0	33.7	17.1	198.9
11 JPN	155.8	0.0	14.9	2.8	13.8	1.4	25.2	12.4	101.4
12 AUS	32.8	30.6	0.0	6.8	2.1	0.1	7.6	0.7	16.9
13 NZL	2.3	3.1	5.0	0.0	0.7	0.0	1.1	0.6	4.9
14 CAN	14.1	11.1	2.1	0.5	0.0	0.7	3.4	4.9	20.8
15 COL	1.1	0.6	0.1	0.0	0.4	0.0	0.2	0.6	10.8
16 GCC	33.3	91.9	1.9	1.2	1.9	0.1	22.4	0.9	70.7
17 MEX	4.1	2.9	0.9	0.1	10.1	3.3	1.7	0.0	17.3
18 ROW	252.4	89.2	14.0	2.2	27.5	7.0	37.0	18.6	442.7
Total	1123.2	708.9	173.9	33.7	413.8	34.8	359.7	245.3	1997.5

Source: Own calculation based on GTAP DB prelease candidate 8.1 (Feb. 2013)

Tab. 5: Bilateral trade by sector between Korea and the EU27 in 2007 (million US$)

Sector	Korea's exports to EU27*	Korea's imports from EU27**	Korea's trade balance with the EU27
1 Paddy rice	0.0130	0.0000	0.0130
2 Wheat	0.0006	0.2078	-0.2071
3 Cereal grains nec	0.0341	0.1756	-0.1415
4 Vegetables, fruit, nuts	5.0764	4.6534	0.4229
5 Oil seeds	0.2460	0.1616	0.0844
6 Sugar cane, sugar beet	0.0021	0.0042	-0.0021
7 Plant-based fibers	0.0768	8.1449	-8.0681
8 Crops nec	9.9676	54.9625	-44.9949
9 Cattle, sheep, goats, horses	0.0125	3.4885	-3.4760
10 Animal products nec	0.3674	51.8350	-51.4675
11 Raw milk	0.0002	0.0453	-0.0451
12 Wool, silk-worm cocoons	0.0344	1.0122	-0.9778
13 Forestry	0.7629	42.8009	-42.0380
14 Fishing	3.6461	0.6634	2.9827
15 Meat: cattle, sheep, goats	0.6100	31.0010	-30.3911
16 Meat products nec	2.3652	554.3784	-552.0132
17 Vegetable oils and fats	0.4981	107.7000	-107.2019
18 Dairy products	1.5997	198.3600	-196.7604
19 Processed rice	2.4457	0.0362	2.4094
20 Sugar	0.0525	0.5350	-0.4825
21 Food products nec	153.0986	915.6642	-762.5656
22 Beverages and tobacco	12.3593	497.1434	-484.7842
23 Mining	7.6201	50.6594	-43.0393
24 Manufacturing sectors	53,315.3398	32,491.9844	20,823.3555
25 Services	11,854.7568	14,172.8623	-2,318.1055
Total	65,370.9859	49,188.4797	16,182.5062

*: Exports are evaluated at FOB prices
**: Imports are evaluated at CIF prices

Source: Own calculation using GTAP DB prelease candidate 8.1 (Feb. 2013)

Scenarios of the EU-Korea FTA

The scenarios for the EU-Korea FTA include baseline scenarios in which the EU-Korea FTA is not implemented and policy scenarios with the EU-Korea FTA being put into effect. The baseline scenarios provide a picture of what the global economy is expected to look like without the EU-Korea FTA, while the policy scenarios are used to examine the impacts of the EU-Korea FTA. The difference between the baseline scenarios and a policy scenario shows the net effects of a policy scenario related to the EU-Korea FTA.

Korea has established 9 free trade agreements with 48 countries so far, while the EU has preferential trade agreements or economic partnership agreements in place with 49 countries. Baseline scenarios should or could include all these free trade agreements that both Korea and the EU implemented. However, for the sake of simplicity, no baseline scenario for the EU is provided considering explicitly all its FTAs but the one concluded with Korea. On the other hand, only the 5 free trade agreements that have been established by Korea since 2007, the base year of the database used in this study, are included in the baseline scenarios: the Korea-ASEAN FTA that became effective in 2007, the Korea-India comprehensive economic partnership agreement (CEPA) that entered into force in 2010, the Korea-Peru FTA that came into effect in 2011, the Korea-US FTA that became effective in 2012 and the Korea-Turkey FTA that entered into force in 2013.

Such baseline scenarios are run and then policy scenarios are conducted using the updated data coming from the baseline scenarios.

The policy scenario for the EU-Korea FTA includes 8 scenarios based on the 8 different levels of regional aggregation of the EU, as shown in Table 2. For example, scenario EU1 is about a free trade agreement between Korea and the 27 member states of the EU represented as one region. Scenario EU7 is about an FTA between Korea and the 27 EU member states aggregated into 7 regions; i.e. into EC6, DIU, G, EP, AFS, CEEC10 and BR. Scenario EU12 is about an FTA between Korea and the EU being represented as 12 regions, which is like EU7 but 'EC6' being disaggregated into its founding members Belgium, France, Germany, Italy, Luxembourg and the Netherlands. scenario EU14 is based on EU12 but 'DIU' being split into Denmark, Ireland and the United Kingdom making a total of 14 regions. Scenario EU15 has 15 regions similar to EU14 with the only difference that 'EP' is disaggregated into Spain and Portugal. Scenario EU17 in comparison to EU15 has 'AFS' disaggregated into Austria, Finland and Sweden. Scenario EU26 represents the EU as 26 regions, in which 'CEEC10' is made up of the countries which joined the EU in 2004; i.e. Czech Republic, Estonia, Hungary,

Latvia, Lithuania, Poland, Slovakia, Slovenia, Cyprus and Malta. Finally, scenario EU27[34] is about an FTA between Korea and the EU in which the latter is represented by all 27 member states individually. This means that 'BR' is split into Bulgaria and Romania.

For each of these 8 policy scenarios, it is assumed that the EU eliminates tariffs on all imports from Korea, except on services. Likewise, for Korea the assumption is made that it eliminates tariffs on 22 commodities imported from the EU, except paddy and processed rice as well as services. 100% cut of tariffs on agricultural goods, except for Korea's paddy and processed rice, and on manufactured commodities is considered for all policy scenarios[35]. Therefore, the effects of the policy scenarios could be regarded as minimum ones of the EU-Korea FTA[36].

Simulation Results

As results of simulating the impact of the EU-Korea FTA real GDP, equivalent variation (EV) as a measure of welfare, total exports and total imports at the macroeconomic level and output by sector at the microeconomic level are presented. Only those of Korea and Germany are shown and discussed due to space limitations[37]. This will be done for the 8 different policy scenarios as elaborated in the previous chapter.

---

34 Please note that the term 'EU27' is used in the text with two different meanings. First, it refers to the European Union with all its (27) member states but Croatia. Second, the text relates to scenario EU27 which is characterized by the fact that each of these 27 member states is represented individually.
35 Applied ad valorem tariff rates of Korea and the EU in Table 3 are cut by 100% for scenario EU1. For scenario EU27 the ad valorem tariff rates applied by Korea on imports from each of the 27 member states of the EU are depicted in Table A2 and those applied by each of the 27 member states of the EU on imports from Korea in Table A3. All these tariff rates are cut completely. For the other scenarios EU7, EU12, EU14, EU15, EU17 and EU26, as seen in Table 2, applied ad valorem tariff rates in Tables A2 and A3 are aggregated according to their regional specification and are also cut by 100%.
36 Besides the reduction of tariffs on primary and manufactured goods, elimination of non-trade barriers (NTBs) such as trade liberalization in services and elimination of barriers to foreign direct investment (FDI) are relevant to the EU-Korea FTA. The consideration of trade liberalization in services and elimination of barriers to FDI would lead to bigger effects of the EU-Korea FTA. See KO (2014a) for the impacts of trade liberalization in services of the EU-Korea FTA and KO (2014b) for the effects of eliminating the barriers to FDI of the EU-Korea FTA.
37 The simulation results of other countries are available from the author on request.

Impacts on Korea

Figure 2 illustrates the effects implementing the EU-Korea FTA on Korea's real GDP. The figure is quite indicative. Firstly, this FTA has a positive impact on this variable. Secondly, the simulations yield that the level of regional aggregation of the EU27 also matters; the lower the level of regional aggregation of the EU27 in the model, the higher the increase in Korea's real GDP –at least it does not decline. This becomes obvious, if one compares the second and all further scenarios with the first one, the EU1. The rise in real GDP is in all of those higher. As a reminder, the level of aggregation declines as one goes across the scenarios from the left to the right as also indicated by the names of the scenario. Especially, the change from scenario EU7 to EU12 is considerable; i.e. if the founding member states are not anymore considered as a single aggregate (EC6 in EU7) but depicted individually in EU12. This leads to an addition of 5 entities in the model. However, the number of additional regions/countries seems to be not as important as is the type of countries. Among those countries individually represented in EU12 are the largest economies of all EU27 member states. Even more important can be that Germany and France are the 1st and 2nd largest trading partners of the EU27, respectively with Korea. This effect may also be seen from Table 6. Due to implementing this FTA, Korea's real GDP goes up 26 percent more in EU12 as compared to EU7.

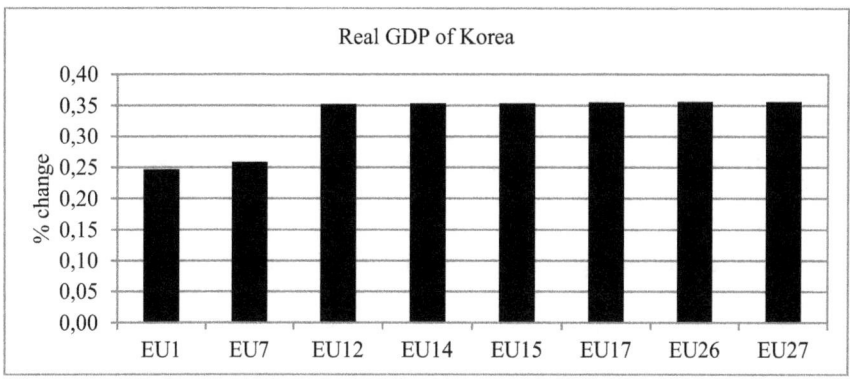

Fig. 2: *Effects of 8 policy scenarios on Korea's real GDP (% change)*

Source: Own calculation

However, the number of entities with which the EU27 is represented does not matter in all cases. This can be seen by comparing policy scenarios EU26 with EU17. In the former the EU27 is modelled using 9 entities more than in the latter because the aggregate CEEC10 is replaced by its individual countries. And yet,

Korea's real GDP goes up only marginally more (0.38 percent) in EU26 relative to EU17 (see Table 6). The level of aggregation has an impact as stated above; i.e. the lower the aggregation, the higher the increase. But the number of entities alone is not a good indicator.

Tab. 6: Effects of 8 policy scenarios on Korea's real GDP

Policy scenario	Changes in real GDP of Korea, in percent	Step-wise difference, in percentage points	Step-wise difference, in percent	Percentage Difference relative to EU1
EU1	0.2464			
EU7	0.2583	0.0119	4.83	4.83
EU12	0.3521	0.0938	36.31	42.90
EU14	0.3532	0.0011	0.31	43.34
EU15	0.3533	0.0001	0.03	43.38
EU17	0.3548	0.0015	0.42	43.99
EU26	0.3558	0.0010	0.28	44.40
EU27	0.3559	0.0001	0.03	44.44

Source: Own calculation

Table 6 provides more detailed information about these effects. For example, Korea's real GDP increases by 0.0011 percentage points more in EU14 (0.3552%) relative to EU12 (0.3521%), which corresponds to an increase which is only 0.31% higher. Please recall that in policy scenario EU14 the region 'DIU' is split into its constituencies Denmark, Ireland and the United Kingdom. Of special interest is a comparison of scenarios EU27 and EU1. Representing the EU27 as 1 entity (EU1) or as 27 ones (EU27) affects the impact on Korea's real GDP when the FTA is put into place. In the latter this variable goes up 44.44% more than in the former, i.e. 0.3559% compared to 0.2464%. This is indicative of the fact that the less aggregation the higher the impact of the EU-Korea FTA on Korea's real GDP.

Figure 3 and Table 7 present the effects of the 8 different policy scenarios for the EU-Korea FTA on Korea's equivalent variation as a measure of welfare. A higher level of regional disaggregation is to lead to a higher rise of Korea's welfare. For example, Korea's welfare increases by $114.7 million from $2.655 billion in case of EU1 to $2.769 billion in case of EU7, which corresponds to a 4.32% rise of Korea's welfare, when the EU as one region is disaggregated into 7 regions. A further regional disaggregation of the EU from EU7 to EU12 leads to a 31.29% increase in Korea's welfare from $2.769 billion in case of EU7 to $3.636 billion in case of EU12.

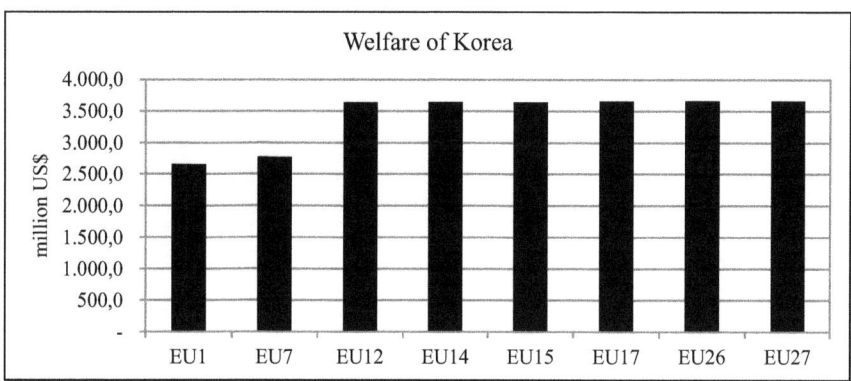

Fig. 3: *Effects of 8 policy scenarios on Korea's welfare (million US$)*
Source: Own calculation

Tab. 7: *Effects of 8 policy scenarios on Korea's welfare*

Policy scenarios	Changes in welfare of Korea (million US$)	Step-wise difference (million US$)	Step-wise difference in percent	Percentage Difference relative to EU
EU1	2,654.6			
EU7	2,769.3	114.7	4.32	4.32
EU12	3,635.9	866.6	31.29	36.97
EU14	3,644.3	8.4	0.23	37.28
EU15	3,645.1	0.8	0.02	37.31
EU17	3,659.5	14.4	0.40	37.86
EU26	3,667.3	7.8	0.21	38.15
EU27	3,667.6	0.3	0.01	38.16

Source: Own calculation

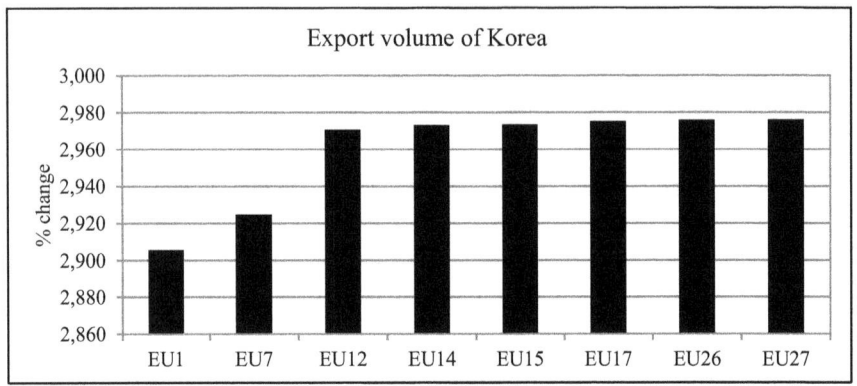

Fig. 4: Effects of 8 policy scenarios on Korea's total export volume (% change)
Source: Own calculation

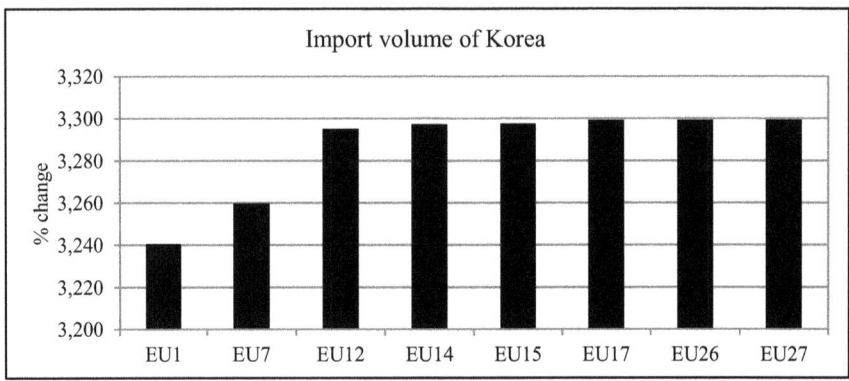

Fig. 5: Effects of 8 policy scenarios on Korea's total import volume (% change)
Source: Own calculation

Further regional splitting of the EU leads to a negligibly slight increase in Korea's welfare. Finally, regional disaggregation of the EU from one region to 27 regions leads to a 38.16% increase in Korea's welfare from $2.655 billion in case of EU1 to $3.668 billion in case of EU27, which indicates that the most disaggregated data of the regions, if possible, should be used for an impact analysis of the EU-Korea FTA, as in the case of its impact on real GDP of Korea.

Tab. 8: Effects of scenarios EU1 and EU27 on Korea's output by sector

Sector (abbreviation)	Change in EU1 in percent	Changes in EU27 in percent	Difference in Change (EU27 – EU1), in percentage points
1 Paddy rice (pdr)	0.14	0.12	-0.03
2 Wheat (wht)	-0.15	-0.19	-0.04
3 Cereal grains nec (gro)	-1.71	-2.55	-0.84
4 Vegetables, fruit and nuts (v_f)	-0.16	-0.28	-0.12
5 Oil seeds (osd)	-0.08	-0.13	-0.04
6 Sugar cane, sugar beet* (c_b)	0.00	0.00	0.00
7 Plant-based fibers (pfb)	0.51	0.65	0.14
8 Crops nec (ocr)	-0.02	-0.11	-0.09
9 Cattle, sheep, goats, horses (ctl)	0.48	0.85	0.37
10 Animal products nec (oap)	-3.43	-2.98	0.45
11 Raw milk (rmk)	-3.75	-3.87	-0.12
12 Wool, silk-worm cocoons (wol)	0.31	0.39	0.08
13 Forestry (frs)	-0.70	-0.93	-0.23
14 Fishing (fsh)	-0.70	-1.03	-0.34
15 Meat: cattle, sheep, goats, horse (cmt)	0.26	0.56	0.30
16 Meat products nec (omt)	-5.09	-4.54	0.54
17 Vegetable oils and fats (vol)	-1.03	-1.32	-0.28
18 Dairy products (mil)	-3.54	-3.53	0.01
19 Processed rice (pcr)	0.08	0.07	-0.01
20 Sugar (sgr)	-1.64	-2.22	-0.58
21 Food products nec (ofd)	-2.55	-3.77	-1.22
22 Beverages and tobacco products (b_t)	0.85	1.29	0.44
23 Mining (MNG)	-0.26	-0.23	0.03
24 Manufacturing sectors (MNF)	1.62	1.67	0.05
25 Services (SRV)	0.13	0.19	0.07

* There is no domestic production of sugar cane and sugar beet in Korea

Source: Own calculation

Figures 4 and 5 depict the effect of the EU-Korea FTA on Korea's total export and import volumes. Changes in Korea's total export volume display quite a similar pattern to those in Korea's total import volume. A higher level of regional disaggregation of the EU is to lead to a higher increase in Korea's export and import volumes. Due to the EU-Korea FTA the increase in Korea's export volume is in scenario EU27 by 2.42% stronger than in EU1 (2.98% compared to 2.91%). The similar number for import volume amounts to 1.82% (3.3% relative to 3.24).

Table 8 shows the effects of scenarios EU1 and EU27 on Korea's domestic production by sector. As can be seen the implementation of the EU-Korea FTA leads to relatively small changes in output by sector and a complete disaggregation of the EU27 into all of its 27 member states results in stronger decreases or smaller increases in output of most agricultural sectors of Korea (negative numbers in the last column of this table). The opposite effect is limited to a few sectors. Nevertheless, the differences between the two scenarios are small and exceed only in the case of sector 21 Food products nec 1 percentage point. Also the increase in production in the aggregate, 24 Manufacturing sectors[38] differs only by 0.02 percentage points. Yet, that increase is large enough to offset in both scenarios the declines in agricultural sectors which explains the rises in Korea's real GDP, as shown in Figure 2 and Table 6 above.

Impacts on Germany

Tables 9 and 10 provide information about the impact of six policy scenarios[39] as regards the EU-Korea FTA on Germany's changes in real GDP and welfare. As can be seen from these numbers differences in the level of regional aggregation of the EU27 do not lead to significantly diverse impacts on Germany's real GDP and welfare since they are equal up to three places after the decimal point: 6 policy scenarios. Therefore, irrespective of the level of regional aggregation of the EU, the EU-Korea FTA leads to an additional increase in Germany's real GDP of 0.025 percent and in its welfare of $1.2 billion.

However, a very close look at Table 9 reveals that, as a result of scenario EU14 and scenario EU26, Germany's real GDP increases more by 0.00007 and 0.00003 percentage points, respectively, than the corresponding preceding sce-

---

38 The share of the manufacturing sector and the services sector in the total value of output at market prices in 2007 is 44.9% and 50.3%, respectively, while that of the primary sector including agriculture is 4.8%.
39 Since Germany is not treated as an individual country in scenarios EU1 and EU7, it is not possible to measure their aggregation effects on this country. Therefore, only 6 policy scenarios are relevant to Germany as an individual country.

nario. It is because the region 'DIU' of scenario EU12 is disaggregated into Denmark, Ireland and the United Kingdom in scenario EU14 and the region 'CEEC10' of scenario EU17 is split into Czech Republic, Estonia, Hungary, Latvia, Lithuania, Poland, Slovakia, Slovenia, Cyprus and Malta in scenario EU26. By contrast, as a result of scenario EU15 and scenario EU17, Germany's real GDP declines by 0.00003 and 0.00003 percentage points, respectively. It is because the region 'EP' of scenario EU14 is disaggregated into Spain and Portugal in scenario EU15 and the region 'AFS' of scenario EU15 is split into Austria, Finland and Sweden in scenario EU17. The disaggregation of the region 'BR' of scenario EU26 into Bulgaria and Romania in scenario EU27 has no impact on Germany's real GDP.

Tab. 9: *Effects of 6 policy scenarios on Germany's real GDP*

Policy scenarios	Changes in real GDP of Germany in percent	Step-wise difference in percentage points	Step-wise difference, in percent	Percentage difference relative to EU12
EU12	0.02473			
EU14	0.02480	0.00007	0.28	0.28
EU15	0.02477	-0.00003	-0.12	0.16
EU17	0.02474	-0.00003	-0.12	0.04
EU26	0.02477	0.00003	0.12	0.16
EU27	0.02477	0.00000	0.00	0.16

Source: Own calculation

Quite a similar pattern of changes in the effects of further regional disaggregation of the EU happens to Germany's welfare as a result of 6 policy scenarios, as seen in Table 10.

In Tables 11 and 12 the impacts of 6 scenarios on Germany's total export and import volumes, respectively are displayed. The different levels of regional aggregation of the EU27 do not influence the impacts on these two types of goods of Germany. The effects of all these 6 scenarios are very similar. They lead to an additional increase in Germany's total export volume of 0.177 percent and in its total import volume of 0.253 percent.

Tab. 10: Effects of 6 policy scenarios on Germany's welfare

Policy scenarios	Changes in welfare of Germany (million US$)	Step-wise difference (million US$)	Step-wise difference in percent	Percentage difference relative to EU12
EU12	1,202.1			
EU14	1,205.2	3.1	0.26	0.26
EU15	1,203.5	-1.7	-0.14	0.12
EU17	1,201.8	-1.7	-0.14	-0.02
EU26	1,203.2	1.4	0.12	0.09
EU27	1,203.2	0.0	0.00	0.09

Tab. 11: Effects of 6 policy scenarios on Germany's total export volume

Policy scenarios	Changes in total export volume in percent	Step-wise difference, in perc. points	Step-wise Difference in percent	Percentage difference relative to EU12
EU12	0.1765			
EU14	0.1768	0.0003	0.17	0.17
EU15	0.1766	-0.0002	-0.11	0.06
EU17	0.1765	-0.0001	-0.06	0.00
EU26	0.1766	0.0001	0.06	0.06
EU27	0.1766	0.0000	0.00	0.06

Tab. 12: Effects of 6 policy scenarios on Germany's total import volume

Policy scenarios	Changes in total import volume, in percent	Step-wise difference in perc. points	Step-wise Difference in percent	Percentage Difference relative to EU12
EU12	0.2525			
EU14	0.2529	0.0004	0.16	0.16
EU15	0.2527	-0.0002	-0.08	0.08
EU17	0.2524	-0.0003	-0.12	-0.04
EU26	0.2525	0.0001	0.04	0.00
EU27	0.2525	0.0000	0.00	0.00

Source: Tab. 10-12 own calculation

EU-South Korea Free Trade Agreement 153

In summary, as far as Germany is concerned, changes in real GDP, total export and total import volumes all follow the same pattern as one goes from scenario EU12 to EU27.

Conclusion

This chapter aims at assessing the economic effects of implementing the EU-Korea FTA that entered into force in July 2011 on the economies of Korea and Germany, in particular, on their agricultural sectors, using a static multi-region, multi-sector CGE model. An extended version of the static GTAP model that incorporates the interaction of trade liberalization and capital accumulation based on classical growth theory and the GTAP database prelease candidate 8.1 which was released in February 2013 are used for this study. The focus of the analyses is on differences in spatial aggregation of the EU27. The base year of the GTAP database prelease candidate 8.1 is 2007. This database includes 134 countries/regions and 57 sectors. In compliance with the research purpose, the 57 sectors are aggregated into 25 sectors with 14 primary sectors, 8 processed food sectors, one aggregated mining sector, one aggregated manufacturing sector and one aggregated services sector. The 134 countries/regions are aggregated into a total of 18 countries/regions. One of those represents the 27 member states of the EU (EU27). Croatia is included in the rest of world (ROW). The spatial aggregation of the EU27 differs. 8 policy scenarios are taken into account ranging from having the EU27 included in the simulation model as one regional entity (EU1) up to considering all its 27 member states individually (EU27).

Simulation results of the EU-Korea FTA are presented in terms of its impact on real GDP, equivalent variation (EV) measuring total welfare, total exports and total imports of Korea and of Germany. In addition, effects on output by sector are also reported.

Major findings of this study are as follows. A higher level of regional disaggregation of the EU is predicted to lead to a higher increase in Korea's real GDP and welfare. Regional disaggregation of the EU from one region (EU1) to 27 countries (EU27) is to result in a 44.44% increase in Korea's real GDP from 0.2464% in case of EU1 to 0.3559% in case of EU27 as well as to a 38.16% increase in Korea's welfare from $2.655 billion in case of EU1 to $3.668 billion in case of EU27. A higher level of regional disaggregation of the EU is to lead to a higher increase in Korea's export and import volumes as well. Regional disaggregation of the EU matters to Korea. These findings indicate that the most disaggregated data of the regions, if possible, should be used for an impact analysis of the EU-Korea FTA.

A further regional disaggregation of the EU is to lead to a further decrease in domestic production of most agricultural sectors of Korea, which implies that Korea has a comparative disadvantage in most agricultural sectors with respect to the EU.

In particular, however, the different level of regional aggregation of the EU has no difference in its impact on the real GDP and welfare of Germany as well as on Germany's total export and import volumes.

The findings of this paper indicate that the most disaggregated data of 27 member states of the EU should be used for an analysis of the impacts of the EU-Korea FTA, in particular, on the Korean economy. It is because all the previous studies on the EU-Korea FTA, i.e. COPENHAGEN ECONOMICS & FRANCOIS (2007), DECREUX, MILNER and PÉRIDY (2010), KIM et al. (2005), KO (2006, 2007, 2014A, 2014B), and KO and LEE (2008) used the EU as an aggregated entity. It is very likely that they underestimated the potential effects of the EU-Korea FTA, in particular, on the Korean economy.

This paper could be improved in three ways. First, one is to use a dynamic CGE model that makes it possible to explicitly take into account different timing of policy implementations related to the tariff reductions agreed upon in the EU-Korea FTA and capital accumulation over time. Second, in addition to the tariff reductions, trade liberalization in services should be considered. To get a more accurate measure of the impacts of the liberalization of trade in services, tariff equivalents of services could be estimated econometrically, for example, using a gravity model. Third, another improvement would be to use a CGE model that explicitly takes into account foreign direct investment (FDI) and foreign commercial presence differentiated by the country of location and ownership. It is because the impacts of liberalization of barriers to FDI and foreign affiliate sales on the Korean economy are predicted to be much bigger than those of trade liberalization through tariff elimination (KO, 2014b). Due to the lack of bilateral data on foreign assets and liabilities, it is assumed in the GTAP model that a global trust collects the savings of all regional households and allocates it to regional investment on their behalf. These improvements will be dealt with in future research.

List of References

Armington, P.S. (1969): A Theory of Demand for Products Distinguished by Place of Production. International Monetary Fund Staff Paper, 16 (1), 159-178.
Baldwin, R.E. (1989): The Growth Effect of 1992. Economic Policy, 9, 247-281.
Baldwin, R.E. (1992): Measurable Dynamic Gains from Trade. Journal of Political Economy, 100, 162-174.
Copenhagen Economics & Francois (2007): Economic Impact of a Potential Free Trade Agreement (FTA) Between the European Union and South Korea. Copenhagen Economics.

Decreux, Y., Milner, C. and Péridy, N. (2010): The Economic Impact of the Free Trade Agreement (FTA) between the European Union and Korea. Report for the European Commission.

Grant, J.H., Hertel, T.W. and Rutherford T.F. (2006): Extending General Equilibrium to the Tariff Line: U.S. Dairy in the Doha Development Agenda. Paper prepared for presentation at the American Agricultural Economics Association Annual Meeting, Long Beach, CA, July 23-26.

GTAP DB (2013): Prelease candidate 8.1.

Hanoch, G. (1975): Production and Demand Models in Direct or Indirect Implicit Additivity. Econometrica, 43, 419.

Hertel, T.W., Peterson, E.B., Preckel, P.V. Surry, Y. and Tsigas, M.E. (1991): Implicit Additivity as a Strategy for Restricting the Parameter Space in CGE Models. Economic and Financial Computing, 1 (1), 265-289.

Hertel, T.W. (ed.) (1997): Global Trade Analysis: Modeling and Applications. Cambridge University Press.

Kim, H., Lee, C., Kim, K., Kang, J. and Park, S. (2005): An Analysis of the Economic Effects of a Korea-EU FTA and its Implications on the Korean Economy (in Korean). KIEP-study 05-09, Korea Institute for International Economic Policy.

Ko, J.-H. (1993): Őkonomische Analyse von Energie- und Volkswirtschaft auf der Basis allgemeiner Gleichgewichtsmodelle. Europäische Hochschulschriften, 1420, Verlag Peter Lang, Frankfurt/Wien/ Paris/New York.

Ko, J.-H. (2006): A study on Economic Effects of a Korea-EU FTA on the Korean Economy by Sector: A Dynamic CGE Approach. Paper presented at a Joint Seminar on FTA Policy of Korea and Korea-EU FTA, Evaluation and Prospect organized by the European Union Studies Association of Korea and the Korean Society of Contemporary European Studies, Graduate School of International Studies, Seoul National University, 25 August.

Ko, J.-H. (2007): Economic Impacts of a Korea-EU FTA: Focusing on Manufacturing Sectors (in Korean). Lee, J.-W., S.-H. Shin, J.-H. Ko, S.-C. Park, J.-H. Jung, H.-B. Chai and C.-R. Bang (eds.). Korea-EU FTA: the Status Quo, Prospects and Strategies (in Korean), Nopi Gipi, Seoul.

Ko, J.-H. (2014a): A Study on the Economic Effects of the Korea-EU FTA Using a Dynamic CGE Model. Journal of International Trade & Commerce, 10 (1), 225-250.

Ko, J.-H. (2014b): Economic Assessment of Korea-EU FTA using a CGE Model with FDI. Journal of European Union Studies, 36, 37-72.

Ko, J.-H. and Lee, J.-W. (2008): A Korea-EU Free Trade Agreement and Analysis of its Economic Effects using a Dynamic CGE Model. The Journal of Contemporary European Studies, 26 (3), 159-186.

Lenzen, M. (2011): Aggregation versus Disaggregation in Input-Output Analysis of the Environment. Economic Systems Research, 23 (1), 73-89.

Pelikan J. and Brockmeier, M. (2008): Methods to Aggregate Import Tariffs and their Impacts on Modeling Results. (https://www.gtap.agecon.purdue.edu/resources/download/3461.pdf).

The European Commission (2013a): Top Trading Partners. (http://trade.ec.europa.eu/doclib/docs/2006/september/tradoc_122530.pdf).

The European Commission (2013b): MEMO - The EU's bilateral trade and investment agreements, where are we? (http://trade.ec.europa.eu/doclib/docs/2012/november/tradoc_ 150 129.pdf).

Zhang, D., Caron, J., Winchester, N. and Karplus, V.J. (2013): Sectoral aggregation bias in the accounting of emissions embodied in trade and consumption. GTAP Resources. (https://www.gtap.agecon.purdue.edu/resources/download/6192.pdf).

Appendix

Tab. A1: Korea's FTAs by country and stage of implementation or negotiation

FTA in effect since	Concluded FTA	Under negotiation	Under review

K-Chile FTA: 1 Apr. 2004
K-Singapore FTA: 2 Mar. 2006
K-EFTA FTA: 1 Sep. 2006
K-ASEAN FTA:
  Agr. in goods: 1 Jun. 2007
  Agr. in services: 1 Jun. 2009
  Agr. in investment: 1 Sep. 2009
K-India CEPA: 1 Jan. 2010
K-EU FTA: 1 July 2011
K-Peru FTA: 1 Aug. 2011
K-U.S. FTA: 15 Mar. 2012
K-Turkey FTA: 1 May 2013
	K-Colombia FTA: 31 Aug. 2012
	K-Australia FTA: 4 Dec. 2013
	K-Canada FTA: 11 Mar. 2014
		K-Indonesia FTA
		K-China FTA
		K-Vietnam FTA
		K-New Zealand FTA
		K-C-J FTA
		RCEP (Regional Comprehensive Economic Partnership) = ASEAN+6
			K-Japan FTA
			K-Mexico FTA
			K-GCC FTA
			K-MERCOSUR FTA
			K-Israel FTA
			K-Malaysia FTA
			K-Central America FTA

- EFTA: European Free Trade Association comprised of Iceland, Norway, Switzerland and Liechtenstein
- ASEAN: Association of Southeast Asian Nations consisting of Brunei Darussalam, Cambodia, Laos, Indonesia, Malaysia, the Philippines, Singapore, Thailand and Vietnam
- EU: European Union consisting of 28 member states such as Austria, Belgium, Bulgaria, Cyprus, Czech Republic, Croatia, Denmark, Estonia, Finland, France, Germany, Greece, Hungary, Ireland, Italy, Latvia, Lithuania, Luxembourg, Malta, Poland, Portugal, Rumania, Slovakia, Slovenia, Spain, Sweden, the Netherlands and the United Kingdom
- GCC: Gulf Cooperation Council or Cooperation Council for the Arab States of the Gulf consisting of Bahrain, Kuwait, Oman, Qatar, Saudi Arabia and the United Arab Emirates
- MERCOSUR: *Mercado Común del Sur (Southern Common Market)* consisting of Argentina, Brazil, Paraguay and Uruguay (Bolivia, Chile, Ecuador and Peru: associate members)
- Central America: Panama, Costa Rica, Guatemala, Honduras, Dominica and El Salvador

Source: Information about Korea's FTAs available at the website of FTA supporting portal (http://www.ftahub.go.kr/kr/)

Tab. A2: Applied ad valorem tariff rates of Korea on imports by sector from 27 member states of the EU (%)

Sector	BEL	FRA	DEU	ITA	LUX	NLD	DNK	IRL	GBR	GR	PRT
pdr	0.0	0.0	0.0	0.0	0.0	0.0	0.0	0.0	0.0	0.0	0.0
wht	0.0	0.0	0.0	0.0	0.0	0.0	0.0	0.0	0.0	0.0	0.0
gro	0.0	0.0	0.0	0.0	0.0	0.0	278.9	0.0	0.0	0.0	0.0
v_f	14.0	91.6	60.6	68.8	0.0	38.7	0.0	0.0	68.9	32.1	0.0
osd	5.0	13.7	8.9	0.0	0.0	0.0	0.0	0.0	0.0	0.0	0.0
c_b	0.0	0.0	0.0	0.0	0.0	0.0	0.0	0.0	0.0	0.0	0.0
pfb	0.0	0.6	0.0	0.0	0.0	0.0	0.0	0.0	0.0	0.0	0.0
ocr	18.4	7.6	19.5	5.1	0.0	5.9	2.0	115.1	18.0	20.0	8.1
ctl	7.2	0.0	9.9	0.0	0.0	7.3	0.0	0.0	0.0	0.0	0.0
oap	3.4	17.6	7.8	4.8	0.0	16.2	3.5	6.1	10.0	8.0	0.0
rmk	0.0	0.0	0.0	0.0	0.0	0.0	0.0	0.0	0.0	0.0	0.0
wol	0.3	4.2	0.0	0.1	0.0	0.0	0.0	0.0	0.1	0.0	0.0
frs	0.1	2.4	0.0	3.1	0.0	1.1	0.1	0.0	6.1	3.0	4.7
fsh	0.0	23.1	10.3	0.0	0.0	26.4	15.8	0.0	27.6	0.0	0.0
cmt	15.0	17.9	11.2	10.8	0.0	17.5	18.0	16.4	17.8	0.0	0.0
omt	25.0	24.9	22.4	27.0	0.0	25.1	25.2	32.9	23.8	0.0	0.0
vol	5.9	12.9	12.7	11.2	0.0	9.6	8.4	0.0	13.9	8.0	8.9
mil	79.3	34.7	51.4	33.7	0.0	37.0	40.1	32.6	17.9	36.0	0.0
pcr	0.0	0.0	0.0	0.0	0.0	0.0	0.0	0.0	0.0	0.0	0.0
sgr	35.1	37.4	14.9	40.0	0.0	40.0	0.0	0.0	8.3	0.0	0.0
ofd	48.9	55.5	177.9	45.4	0.0	80.1	69.6	51.5	30.0	29.2	9.5
b_t	30.3	15.5	36.5	15.3	15.0	27.1	23.2	28.2	20.4	35.0	13.8
MNG	1.0	3.1	2.2	3.1	0.0	1.2	0.7	0.1	1.1	3.2	3.0
MNF	5.2	5.7	6.0	7.0	5.3	2.8	6.5	1.9	4.8	3.6	5.5
SRV	0.0	0.0	0.0	0.0	0.0	0.0	0.0	0.0	0.0	0.0	0.0

Tab. A2: (continue) Applied ad valorem tariff rates of Korea on imports by sector from 27 member states of the EU (%)

Sector	ESP	AUT	FIN	SWE	CYP	CZE	EST	HUN	LVA	LTU	MLT
pdr	0.0	0.0	0.0	0.0	0.0	0.0	0.0	0.0	0.0	0.0	0.0
wht	0.0	0.0	0.0	0.0	0.0	0.0	0.0	0.0	0.0	0.0	0.0
gro	0.0	0.0	55.9	0.0	0.0	0.0	0.0	0.0	0.0	0.0	0.0
v_f	46.3	0.0	0.0	0.0	304.0	47.2	0.0	0.0	0.0	0.0	0.0
osd	0.0	0.0	0.0	0.0	0.0	0.0	0.0	0.0	0.0	0.0	0.0
c_b	0.0	0.0	0.0	0.0	0.0	0.0	0.0	0.0	0.0	0.0	0.0
pfb	0.0	0.0	0.0	0.0	0.0	0.0	0.0	0.0	0.0	0.0	0.0
ocr	30.0	25.8	0.0	0.0	0.0	1.7	0.0	11.5	0.0	0.0	0.0
ctl	0.0	0.0	0.0	0.0	0.0	0.0	0.0	0.0	0.0	0.0	0.0
oap	7.7	13.8	3.1	12.4	0.0	4.3	0.0	7.0	0.0	0.0	0.0
rmk	0.0	0.0	0.0	0.0	0.0	0.0	0.0	0.0	0.0	0.0	0.0
wol	0.0	0.0	0.0	0.0	0.0	0.0	0.0	0.0	0.0	0.0	0.0
frs	1.2	0.4	0.0	0.0	0.0	0.0	0.0	0.7	0.0	0.0	0.0
fsh	0.0	0.0	0.0	0.0	0.0	9.8	0.0	0.0	0.0	0.0	20.0
cmt	15.6	15.2	18.0	17.8	0.0	0.0	0.0	16.9	0.0	0.0	0.0
omt	24.4	25.0	24.9	25.5	0.0	62.9	0.0	25.0	0.0	0.0	0.0
vol	10.3	11.5	12.3	14.5	0.0	0.0	0.0	0.0	0.0	0.0	0.0
mil	35.6	72.3	49.6	0.0	0.0	16.3	0.0	0.0	0.0	47.7	0.0
pcr	0.0	0.0	0.0	0.0	0.0	0.0	0.0	0.0	0.0	0.0	0.0
sgr	0.0	0.0	0.0	40.0	0.0	0.0	0.0	0.0	0.0	0.0	0.0

ofd	27.0	140.5	29.4	49.6	0.0	156.6	69.6	14.7	11.8	11.7	9.9
b_t	14.4	23.1	191.2	20.0	0.0	48.0	0.0	15.0	0.0	0.0	15.0
MNG	3.1	2.8	3.4	0.6	0.0	0.5	4.2	0.5	1.0	1.0	0.0
MNF	5.9	5.8	5.1	5.3	2.1	5.9	3.3	5.5	6.9	6.5	0.5
SRV	0.0	0.0	0.0	0.0	0.0	0.0	0.0	0.0	0.0	0.0	0.0

Tab. A2: *(continue) Applied ad valorem tariff rates of Korea on imports by sector from 27 member states of the EU (%)*

Sector	POL	SVK	SVN	BGR	ROU
pdr	0.0	0.0	0.0	0.0	0.0
wht	0.0	0.0	0.0	0.0	0.0
gro	0.0	0.0	0.0	0.0	0.0
v_f	0.0	0.0	0.0	0.0	0.0
osd	0.0	0.0	0.0	25.0	0.0
c_b	0.0	0.0	0.0	0.0	0.0
pfb	0.0	0.0	0.0	0.0	0.0
ocr	8.0	0.0	0.0	19.8	35.8
ctl	0.0	0.0	0.0	0.0	0.0
oap	10.7	2.5	112.9	0.0	0.0
rmk	0.0	0.0	0.0	0.0	0.0
wol	0.0	0.0	0.0	0.0	0.0
frs	0.0	5.0	0.0	0.0	0.0
fsh	0.0	0.0	0.0	10.4	0.0
cmt	17.7	22.2	0.0	0.0	31.3
omt	25.0	25.5	0.0	27.8	0.0
vol	26.7	0.0	0.0	0.0	0.0
mil	33.4	0.0	0.0	58.9	0.0
pcr	0.0	0.0	0.0	0.0	0.0
sgr	0.0	0.0	0.0	0.0	0.0
ofd	53.8	0.0	8.5	20.7	8.0
b_t	36.1	8.0	15.0	13.9	15.2
MNG	0.0	3.2	0.0	0.0	0.0
MNF	6.2	5.7	6.3	2.2	5.2
SRV	0.0	0.0	0.0	0.0	0.0

Source: Own calculation with GTAP database prelease candidate 8.1 (February 2013)

Tab. A3: Applied ad valorem tariff rates of the 27 EU member states on imports by sector from Korea (%)

Sector	BEL	FRA	DEU	ITA	LUX	NLD	DNK	IRL	GBR	GR	PRT
pdr	0.0	0.0	0.0	0.0	0.0	0.0	0.0	0.0	0.0	0.0	0.0
wht	0.0	0.0	0.0	0.0	0.0	0.0	0.0	0.0	0.0	0.0	0.0
gro	0.0	0.0	20.2	0.0	0.0	21.1	0.0	0.0	5.9	0.0	0.0
v_f	0.0	7.0	9.8	0.0	0.0	7.1	0.0	12.8	7.5	0.0	0.0
osd	0.0	0.0	0.0	0.0	0.0	0.0	0.0	0.0	0.0	0.0	0.0
c_b	0.0	0.0	0.0	0.0	0.0	0.0	0.0	0.0	0.0	0.0	0.0
pfb	0.0	0.0	0.0	0.0	0.0	0.0	0.0	0.0	0.0	0.0	0.0
ocr	12.6	9.8	1.7	2.1	0.0	5.0	5.6	1.8	3.3	14.5	0.6
ctl	0.0	0.0	0.0	0.0	0.0	0.0	0.0	0.0	0.0	0.0	0.0
oap	0.6	0.0	12.5	0.0	0.0	0.0	0.0	0.0	0.6	0.0	0.0
rmk	0.0	0.0	0.0	0.0	0.0	0.0	0.0	0.0	0.0	0.0	0.0
wol	0.0	0.0	0.0	0.0	0.0	0.0	0.0	0.0	0.0	0.0	0.0
frs	0.0	0.0	0.0	0.0	0.0	0.0	0.0	0.0	0.0	0.0	0.0
fsh	3.3	0.7	1.8	0.0	0.0	1.0	0.0	0.0	2.7	0.0	0.0
cmt	0.0	0.0	0.0	0.0	0.0	0.0	0.0	0.0	0.0	0.0	0.0
omt	0.0	7.2	10.1	9.5	0.0	0.0	0.0	0.0	9.7	0.0	0.0
vol	0.0	8.4	11.5	0.0	0.0	0.0	0.0	0.0	9.3	0.0	0.0
mil	0.0	0.0	23.0	0.0	0.0	21.7	0.0	0.0	24.2	0.0	9.5
pcr	0.0	0.0	0.0	0.0	0.0	24.1	0.0	0.0	24.1	0.0	0.0
sgr	0.0	0.0	0.0	0.0	0.0	0.0	0.0	42.7	0.0	0.0	0.0
ofd	9.9	8.0	11.3	8.0	0.0	14.8	4.7	15.7	14.8	8.6	15.7
b_t	4.8	13.5	27.2	1.5	0.0	8.6	0.0	2.1	4.4	0.0	10.8
MNG	0.0	0.0	0.0	0.1	0.0	0.0	0.0	0.0	0.0	0.0	0.0
MNF	4.8	4.8	3.2	3.6	2.0	4.0	4.7	3.6	2.3	2.4	3.2
SRV	0.0	0.0	0.0	0.0	0.0	0.0	0.0	0.0	0.0	0.0	0.0

Tab. A3: (continue) Applied ad valorem tariff rates of the 27 EU member states on imports by sector from Korea (%)

Sector	ESP	AUT	FIN	SWE	CYP	CZE	EST	HUN	LVA	LTU	MLT
pdr	0.0	87.9	0.0	0.0	0.0	0.0	0.0	0.0	0.0	0.0	0.0
wht	0.0	0.0	0.0	0.0	0.0	0.0	0.0	0.0	0.0	0.0	0.0
gro	0.0	0.0	0.0	0.0	0.0	0.0	0.0	0.0	0.0	0.0	0.0
v_f	0.0	10.0	0.0	0.0	0.0	0.0	0.0	0.0	0.0	0.0	0.0
osd	0.0	0.0	0.0	0.0	0.0	0.0	0.0	0.0	0.0	0.0	0.0
c_b	0.0	0.0	0.0	0.0	0.0	0.0	0.0	0.0	0.0	0.0	0.0
pfb	0.0	0.0	0.0	0.0	0.0	0.0	0.0	0.0	0.0	0.0	0.0
ocr	7.8	1.4	0.0	1.0	0.0	5.3	0.0	0.0	0.0	0.0	0.0
ctl	0.0	0.0	0.0	0.0	0.0	0.0	0.0	0.0	0.0	0.0	0.0
oap	0.0	0.0	0.0	0.0	0.0	0.0	0.0	0.0	0.0	0.0	0.0
rmk	0.0	0.0	0.0	0.0	0.0	0.0	0.0	0.0	0.0	0.0	0.0
wol	0.0	0.0	0.0	0.0	0.0	0.0	0.0	0.0	0.0	0.0	0.0
frs	0.0	3.4	0.0	0.0	0.0	0.0	0.0	0.0	0.0	0.0	0.0
fsh	11.0	0.4	0.0	0.0	0.0	1.2	0.0	0.0	0.0	0.0	0.0
cmt	0.0	0.0	0.0	1.6	0.0	0.0	0.0	0.0	0.0	0.0	0.0
omt	31.5	0.0	0.0	0.0	0.0	0.0	16.7	0.0	0.0	0.0	0.0
vol	13.1	0.0	0.0	0.0	0.0	5.3	0.0	0.0	0.0	0.0	0.0
mil	0.0	0.0	0.0	0.0	0.0	0.0	0.0	0.0	0.0	0.0	0.0
pcr	0.0	0.0	0.0	0.0	0.0	0.0	0.0	0.0	0.0	0.0	0.0
sgr	0.0	0.0	0.0	0.0	0.0	0.0	0.0	0.0	0.0	0.0	0.0

EU-South Korea Free Trade Agreement 161

ofd	8.5	12.9	4.4	15.4	17.1	12.1	21.4	7.1	11.2	9.0	5.6
b_t	14.0	6.6	14.0	14.5	14.6	10.3	16.3	16.3	13.7	16.3	31.0
MNG	0.0	0.0	0.0	0.0	0.0	0.3	0.0	0.7	0.0	0.0	0.0
MNF	5.4	5.7	2.6	3.8	5.6	3.6	4.5	2.6	3.0	3.1	1.3
SRV	0.0	0.0	0.0	0.0	0.0	0.0	0.0	0.0	0.0	0.0	0.0

Tab. A3: (continue) Applied ad valorem tariff rates of the 27 EU member states on imports by sector from Korea (%)

Sector	POL	SVK	SVN	BGR	ROU
pdr	0.0	0.0	0.0	0.0	0.0
wht	0.0	0.0	0.0	0.0	0.0
gro	0.0	0.0	0.0	0.0	0.0
v_f	0.0	0.0	0.0	0.0	0.0
osd	0.0	0.0	0.0	0.0	0.0
c_b	0.0	0.0	0.0	0.0	0.0
pfb	0.0	0.0	0.0	0.0	0.0
ocr	12.0	0.0	0.0	0.0	0.0
ctl	0.0	0.0	0.0	0.0	0.0
oap	0.0	0.0	0.0	0.0	0.0
rmk	0.0	0.0	0.0	0.0	0.0
wol	0.0	0.0	0.0	0.0	0.0
frs	0.0	0.0	0.0	0.0	0.0
fsh	0.0	0.0	0.0	0.0	0.0
cmt	0.0	0.0	0.0	0.0	0.0
omt	0.0	0.0	0.0	0.0	7.3
vol	11.4	0.0	0.0	0.0	0.0
mil	0.0	0.0	0.0	0.0	0.0
pcr	24.1	0.0	0.0	0.0	0.0
sgr	0.0	0.0	0.0	0.0	0.0
ofd	16.3	8.4	0.0	18.2	14.3
b_t	17.7	8.9	0.0	24.7	0.0
MNG	0.0	0.0	0.0	0.0	0.1
MNF	3.1	3.0	5.0	5.3	4.2
SRV	0.0	0.0	0.0	0.0	0.0

Source: Own calculation with GTAP database prelease candidate 8.1 (February 2013)

Food Security and WTO Domestic Support Disciplines post-Bali

Alan MATTHEWS

Abstract

The consistency of WTO rules and disciplines with the policy environment needed in developing countries to pursue their food security objectives has long been a source of controversy. At the 9th WTO Ministerial Conference in Bali, Indonesia in December 2013, the G-33 group of developing countries placed the treatment of procurement for public stock-holding for food security purposes on the Bali agenda. Disagreement over how to resolve this issue came close to collapsing the Bali meeting. This chapter discusses the background to the controversy over accounting for producer support when a government implements a public stock-holding scheme for food security purposes, describes the interim mechanism that was agreed at Bali and reviews some possible options for the permanent solution that WTO members have committed to agree before the 11th Ministerial Conference in 2017. The premise of the chapter is that adapting WTO rules on agricultural policies where this can be shown to be justified to enable developing countries to pursue their food security objectives is a preferable approach to simply increasing their limits on trade-distorting support. Two proposals which deserve further consideration in this context are discussed. The first would make explicit allowance in the WTO rules for countries to adjust their measured domestic support for excessive rates of inflation. The second would make a distinction between the use of administered prices for price support and as a safety net in the procurement of food grains for public stock-holding policies for food security purposes. These further amendments could sufficiently adapt the domestic support disciplines in the WTO Agreement on Agriculture to address developing countries' remaining concerns about their ability to pursue their food security goals.

Acknowledgements

I would like to thank Lars BRINK for helpful discussions on some of the issues discussed in this chapter, and Eugenio DIAZ-BONILLA, Jonathan HEPBURN and Christophe BELLMAN and Monika HARTMANN for constructive comments. Any remaining errors are my own.

Introduction

The consistency of WTO rules and disciplines with the policy environment needed in developing countries to pursue their food security objectives has long been a source of controversy. Although food security is recognised in the preamble to the WTO Agreement on Agriculture (AoA) as a non-trade concern which must be taken into account in the reform process to establish a fair and market-oriented agricultural trading system, various commentators as well as developing countries claim that this is not the case or, at least, that it has been inadequately recognised (DE SCHUTTER, 2009; DÍAZ-BONILLA and RON, 2010; GONZALEZ, 2002; HÄBERLI, 2010). Criticisms range from arguments that the AoA rules are lop-sided and favour developed countries in allowing them to continue to heavily support their agricultural sectors, that they unduly constrain the ability of developing countries to pursue their agricultural development and food security policies, and even that they undermine the right to food of developing countries.

Against this background, food security concerns have played an important role in the Doha Round negotiations to revise the AoA (MATTHEWS, 2012a, 2012b). Developing countries sought exemptions from tariff reductions for products they saw as important for their food security, as well as for the right to protect themselves from destabilising import competition. In a June 2000 submission to the WTO Committee on Agriculture, eleven developing countries suggested extending special and differential treatment to allow developing countries greater flexibility to tackle food security and to protect the rural poor (WTO, 2000a). Their concerns were reflected in the WTO General Council Decision of 1 August 2004 (the Framework Agreement) which stated that developing country members "must be able to pursue agricultural policies that are supportive of their development goals, poverty reduction strategies, food security and livelihood concerns" (WTO, 2004a). It went on to specify that "developing country Members will have the flexibility to designate an appropriate number of products as Special Products, based on criteria of food security, livelihood security and rural development needs. These products will be eligible for more flexible treatment." The Framework Agreement further states that a "Special Safeguard Mechanism will be established for use by developing country Members." The failure to reach agreement on revised modalities in December 2008 has been blamed, in part, on disagreements over the design of the Special Safeguard Mechanism which many developing countries saw as an essential instrument to underpin their food security (WOLFE, 2010).

Following the failure to agree on a set of revised modalities in 2008, the Doha Round negotiations appeared to have stalled. The latest effort to break the stalemate took place in the run up to the 9th WTO Ministerial Conference held in Bali in December 2013 when members were asked to agree on a 'mini-package' of issues seen as a down-payment to try to build momentum for the completion of the broader Doha agenda. The origin of this idea was a plan by the WTO Director-General Pascal LAMY for an early harvest of deliverables with a particular focus on the least developed countries (LDCs) in the months preceding the 8th Ministerial Conference in 2011. While this proposal failed to find support, feedback from the negotiating group Chairs and from members indicated that a small package for the 9th Ministerial Conference in Bali built around trade facilitation, an element on agriculture, and an element on development/LDCs might be feasible (MAT-THEWS, 2013).

Three submissions were eventually made as the agricultural elements of the mini-package (in addition, cotton was addressed as part of the LDC element). Two of these were put forward by the G-20 group of developing countries; many of these countries are exporters and have offensive interests in the negotiations.[1] Their proposals were concerned with ensuring the maximum utilisation of tariff rate quotas and making an advance down-payment on the elimination of export subsidies. The other submission was put forward by the G-33 group of developing countries which generally holds rather defensive positions in the negotiations and seeks to maximise the policy space that developing countries have with respect to the use of agricultural policies. Their submission was to advance two measures included in the 2008 revised draft modalities intended to facilitate developing countries in addressing food security issues (BELLMAN et al., 2013). One measure was to add a range of schemes primarily used by developing countries – such as farmer settlement, land reform and other programmes to promote rural development and poverty alleviation – to the general government services classed as green box expenditure which is exempt from disciplines on the overall amount of spending. The other measure would exempt schemes of food purchases at administered prices from low-income or resource-poor producers for food security purposes from counting towards a developing country's maximum permitted ceiling on trade-distorting support.

The suggestion to disregard purchases at official prices for public stocks for food security purposes when calculating a country's total amount of trade-distorting support was particularly contentious; disagreement over this issue nearly derailed the Bali package. In the earlier negotiations in Geneva, India had apparently agreed to a temporary 'peace clause' which would have prevented challenges to schemes for public food security stocks even where procurement took place at minimum official prices. However, widespread opposition in India and elsewhere to the temporary nature of this protection led to India strengthening its stand just before the Bali meeting (Wall Street Journal, 2013). In the end, the conference extended into a further day to allow agreement to be reached.[2] Ministers agreed to an interim mechanism while they worked to find a permanent solution for the issue of public food security stocks for adoption by the 11th Ministerial Conference in 2017.

---

1 The members of the G-20 and G-33 as well as other country groups in the WTO are set out in the WTO website http://www.wto.org/english/tratop_e/dda_e/meet08_ brief08_ e. htm.

2 The extension was also required to deal with Cuba's concern to include language in the trade facilitation text to address the US embargo on Cuban goods, see ICSTD (2013).

This chapter discusses the background at Bali to the controversy over accounting for producer support when a government implements a public stock-holding scheme for food security purposes, describes the interim mechanism that was agreed and reviews some possible options for a permanent solution. The work programme to be undertaken in the Committee on Agriculture will be based on members' existing and future submissions. There is thus potentially an opportunity to take up more broadly the relationship between WTO rules and food security, and to assess the extent to which revisions to these rules are desirable to permit countries to pursue their food security objectives without damaging the food security ambitions of other members. This agenda covers all three pillars of the AoA, including market access, domestic support and export competition, but the work programme will address the domestic support pillar which was at the centre of the Bali controversy (for the market access issues including Special Products and the Special Safeguard Mechanism, see MATTHEWS (2012a), and for a discussion on export restrictions, see ANANIA (2013)).

Important issues are at stake. Domestic support, including market price support, has been rising rapidly in a number of the more advanced developing countries. Some fear that providing more policy space to developing countries runs counter to the expressed objective of the WTO AoA to encourage progressive reductions in agricultural support and protection over time. Some take the normative view that if WTO rules stand in the way of food security, then the rules should be changed (DE SCHUTTER, 2011). Rule changes in the WTO are the outcome of a bargaining process among members. There is no consensus on whether particular rule changes are necessary to promote food security. Increasing the flexibility of some countries to provide support to their farmers is opposed by those for whom this means greater competition in the future. Even if it were agreed that developing countries should be granted greater scope to use, for example, price support measures, there is no obvious standard to suggest where the new limits should be drawn. This chapter does not try to prejudge the outcome of the negotiations in Geneva on a permanent solution. The more modest objective is to evaluate those changes that have been suggested to address some developing country concerns that current AoA domestic support rules are compromising their food security.

Public stock-holding for food security purposes

The G-33 proposal to exclude purchases at official prices for public stock-holding programmes for food security purposes from a country's Current Total Aggregate Measure of Support (AMS) was the most controversial element of the Bali agricultural mini-package. Indeed, in the weeks up to the Ministerial Conference, it seemed that disagreement on this issue threatened to derail an overall agreement. There are two relevant food security provisions in Annex 2 of the AoA. They are the rules on public stockholding for food security purposes (paragraph 3) and the provision of domestic food aid (paragraph 4).

Public stockholding for food security purposes covers "expenditures (or revenue foregone) in relation to the accumulation and holding stocks of products which form an integral part of a food security programme identified in the national legislation". This can include government assistance to private storage of products as part of a food security programme. All such operations have to be conducted subject to three conditions: (i) the volume and accumulation of such stocks has to correspond to predetermined targets in relation solely to food security; (ii) the processes of stock accumulation and disposal have to be financially transparent; and (iii) food purchases by the government have to be made at current market prices and sales from food security stocks have to be made at no less than the prevailing domestic market price for the product and quality in question. However, programmes in developing countries which provide food at subsidised prices with the objective of meeting the food requirements of the urban and rural poor are assumed to be in conformity with this paragraph. A further footnote allows programmes in developing countries to be considered in conformity with all of the above conditions (i), (ii) and (iii) simply if their operation is transparent and conducted in accordance with official published objective criteria or guidelines, i.e. a less stringent set of criteria than those required for developed countries. However, if the stocks of foodstuffs are acquired and released at administered prices, the difference between the acquisition price and the external reference price will have to be accounted for in the product's AMS.

Domestic food aid is also included under the green box. It is defined as expenditures (or revenue foregone) in relation to the provision of domestic food aid to sections of the population in need. Eligibility to receive the food aid should be subject to clearly-defined criteria related to nutritional objectives. Such aid should be in the form of direct provision of food to those concerned or the provision of means to allow eligible recipients to buy food either at market or at subsidised prices. Food purchases by the government should be made at current market prices and the financing and administration of the aid should be transparent.

The G–33 argued that requiring the inclusion of public stocks purchased at administered prices in the product's AMS means that several developing countries are in danger of reaching or exceeding their permitted limits. Specifically, the G-33 wanted to fast-track amendments in the revised draft modalities that would allow developing country governments to buy food from low-income or resource-poor producers at government-set prices (therefore providing price support for producers) with the objective of stocking it for food security purposes or distributing it as food aid – without having to count a price difference in the product's AMS, which is subject to limits.

The draft modalities would add a sentence to the footnote to Annex 2, paragraph 3 dealing with public stock-holding that: "Acquisition of stocks of foodstuffs by developing country Members with the objective of supporting low-income or resource-poor producers shall not be required to be accounted for in the AMS". A further amendment would allow "the acquisition of foodstuffs at subsidised prices when procured generally from low-income or resource-poor producers in developing countries with the objective of fighting hunger and rural poverty" to be deemed to be in conformity with the conditions for public food stocks to be eligible for the green box (WTO, 2008).

Food reserves can play an important role in developing countries faced with volatility in both food availability and food prices, and food assistance programmes provide a vital safety net for food-insecure families. Hence it is important and appropriate that current AoA rules recognise that the creation of food reserves and the provision of domestic food aid which meet the specified conditions should not be restricted. If food for public reserves or food assistance programmes is purchased at market prices, then these programmes qualify as green box programmes without restriction.

However, the rationale of the G-33 proposal is that these programmes should also be able to provide price support to producers, or at least low-income and resource-poor producers without this price support being required to count towards the product's AMS (South Centre, 2013). It is not surprising that the suggestion received a mixed reception given that it would breach a basic criterion for green box programmes that they should not provide price support to producers. Countries critical of the G-33 plan focused on the systemic impact of changing the current rules to such an extent outside of a wider negotiation. They also highlighted the potential trade-distorting consequences of any such change. Developed countries in particular expressed concern that the move could allow countries to provide unlimited sums of trade-distorting farm support to their farmers – potentially undermining producers in other countries (ORYZA, 2013).

In the weeks leading up to the Bali Conference, negotiations focused on a possible 'peace clause' that would provide some additional flexibility for specific members on the basis that this would be time limited, non-automatic, and create no or minimal trade or production distortions. The intention was to provide some additional breathing space for members having trouble respecting their commitments in respect of public food security stocks while working to find a more lasting solution. But key differences remained until the last moment, including

Food Security and WTO Domestic Support Disciplines post-Bali   169

whether the flexibility delivered under such a mechanism should be automatic or non-automatic, and on what would happen when the peace clause expired.

At Bali, WTO members agreed on an interim solution, and committed to negotiate on an agreement for a permanent solution for the issue of public stockholding for food security purposes for adoption by the 11th Ministerial Conference in 2017 (WTO, 2013a). The Ministerial Decision declared that, as an interim solution, WTO members shall refrain from challenging through the WTO dispute settlement mechanism compliance of a developing member with its Total AMS or *de minimis* AMS limits in relation to support provided for traditional staple food crops in pursuance of public stockholding programmes for food security purposes existing as of the date of this Decision, provided it complies with a number of conditions set out in the Decision.

The conditions relate to notification and transparency requirements, anti-circumvention and safeguards, consultation and monitoring. A WTO member which wants to benefit from this Decision must have notified the Committee on Agriculture that it is exceeding or is at risk of exceeding either or both (sic) of its AMS limits (the Bound Total AMS or the *de minimis* limits) as a result of a stockholding programme for food security purposes.[3] The country must also have fulfilled its domestic support notification requirements under the AoA by ensuring that its notifications for the previous five years are up-to-date. It is required to provide detailed information, on an annual basis, on each public stockholding programme that it maintains for food security purposes as well as relevant statistical information in a format which is set out in an annex to the Decision. Stocks procured under such programmes must not distort trade or adversely affect the food security of other WTO members. Finally, a developing country benefiting from the Decision shall upon request hold consultations with other WTO members on the operation of any of its public stockholding programmes.

Some of the key elements in this Decision are worth underlining. First, it does not address the issue of domestic food aid, where the G-33 proposal would similarly have exempted the acquisition of any foodstuff at subsidised prices when procured generally from low-income or resource-poor producers from a product's AMS. Second, support does not have to be limited to low-income or resource-poor farmers; indeed, market price support by definition will benefit all farms. Third, it only covers stock-holding programmes dealing with staple food crops. Fourth, it only applies to public stockholding programmes for food security purposes existing as of the date of the Decision. Thus it does not provide carte blanche

---

3 The implication that a country can have both a Bound Total AMS and de minimis limits seems to be a drafting error and probably reflects the rather chaotic circumstances in which the Decision was drafted.

for developing countries to initiate new price support programmes linked to public stock-holding activities. Fifth, beneficiary countries are subject to on-going provision of information to allow WTO members to monitor each country's situation. For some countries this will mean a dramatic change in their notification practices; India, for example, had only submitted domestic support notifications up to 2003 by 13 September 2013 (WTO, 2013b). Sixth, the Decision only protects eligible countries from challenge to their commitments under the Agreement on Agriculture. Support schemes can still be challenged under the Subsidies and Countervailing Measures Agreement if such schemes result in adverse effects for third countries.

The period of validity of the Decision was a particularly vexed issue. The original suggestion hammered out in Geneva in the weeks prior to the Conference was that the interim solution would take the form of a "peace clause" with a limited validity of four years, after which the protection of the Decision would become void. India, however, on behalf of the G-33 insisted that the interim solution would remain in place until a permanent solution was found. In the debate during the weeks leading up to the Bali agreement, the provision that a country had to admit that it was likely to breach its AMS commitments in order to benefit from the protection of the "peace clause" was dubbed a "Trojan Horse" by some commentators. They feared that, if a country invoked the Decision, then in four years' time its effective admission that it had violated the AoA would make it a sitting target for a complaint in dispute settlement (HOWSE, 2013). The final text avoids this outcome by adopting the Indian position. Either option would change the dynamics of the negotiations around a permanent solution. In the original suggestion, India, as the demander and a likely beneficiary of the Decision, would have come under strong pressure to reach an agreement, even an unsatisfactory one, before the expiry of the Decision. Now it can negotiate from a stronger position, knowing that its programmes are protected from challenge until a permanent solution is found.

Table 1 shows the developing countries that potentially might benefit from this Decision. These are the countries which have reported making use of public stock-holding for food security purposes at any time since 1995 (for the reasons mentioned in the footnote to the table, this may be an incomplete list because of delays in notifications to the WTO). Some of these countries did not report programme expenditures in recent years, either because a notification was not made (India, Kenya, Sri Lanka) or it was made and no expenditure was reported (Costa Rica, Pakistan, Philippines, South Africa).

Tab. 1: *Expenditure on public stock-holding schemes for food security purposes in the period 2005-2010 by WTO developing country members which had notified such expenditure at least once since 1995*

Member Name	Currency Unit	2005	2006	2007	2008	2009	2010
Botswana	P million	6.4	4.7	3.2	4.3	6.8	-
Brazil	US$ thousand	147,932	156,739	180,941	234,159	236,785	
China	Y million	44,087	50,378	54,200	57,932		
Costa Rica	US$ thousand	-	-	-	-	-	-
India	US$ million						
Indonesia	Rp billion	847	1,078	1,225	698		
Israel	US$ thousand	15,401	9,715	8,512	12,046	12,932	12,590
Kenya	K Sh million						
Kingdom of Saudi Arabia	SRl million			32.6	35.8	24.3	10.2	16.6
Korea, Republic of	W billion	169.8	143.6	163.8	137.1		
Namibia	N$ million	-	-	10.0	5.0	9.0	
Nepal	NPR thousand	260,000		260,000		354,000	67,761
Pakistan	PRs million	-	-				
Philippines	P thousand						
South Africa	R thousand	-	-	-	-	-	-
Sri Lanka	SL Re million						

Note: A '-' symbol in a cell indicates that a notification was received from that country in that year but no expenditure on public stock-holding for food security purposes was reported. A blank cell means that no notification has yet been received from that country, so it is possible that the country made use of a public stock-holding scheme in those years. Burundi, Cambodia and Mali reported expenditure on public stock-holding for food security purposes in their supporting tables used to derive their schedule of commitments but did not report expenditure in any notification since 1995. There may also be countries which never reported expenditure on public stock-holding schemes in any notification but have begun to make use of this measure in recent years but have not yet notified it. For example, Thailand introduced a paddy rice pledging scheme in the 2011/2012 marketing year but it is not yet clear if it will notify this scheme as a public stock-holding scheme for food security purposes. Thailand has not notified its domestic support since 2007 as of January 2014.

Source: Own calculation based on WTO domestic support dataset: Table DS:1 and the relevant supporting tables, cut-off date 28 February 2013, downloaed from http://www.wto.org/english/tratop_e/agric_e/transparency_toolkit_e.htm

It appears that only a small handful of developing countries are potential beneficiaries of the interim solution, which is confined to countries which are implementing public stock-holding schemes for food security purposes at the date of the decision. Even fewer of these countries are likely to need the exemption introduced by the Decision, either because food is procured at market prices or because their Total AMS or *de minimis* limits are not restraining. Given the notification and reporting obligations associated with the Decision, it would not be surprising if India turns out to be the only country which makes use of it during the next four years. The practical consequence of the interim solution will therefore be very limited.

Towards a permanent solution on public stock-holding

The Ministerial Decision recognises that the interim solution is a stop-gap measure and it establishes a work programme to be undertaken in the Committee on Agriculture with the aim of making recommendations for a permanent solution. This work programme will take into account suggestions already made by members as well as future submissions. It is intended that a permanent solution will be ready for approval no later than the 11th Ministerial Conference in 2017. The permanent solution will apply to all developing country members, and not only to those with public stock-holding programmes for food security purposes already in place.

A number of possible options have already been put on the table. One option would be to roll-over the interim solution and extend it to all developing countries by an interpretation that purchases at administered prices for the purposes of public stock-holding for food security purposes would not be deemed to be price support and would not be required to be included in a product's AMS (Option 1). Another option suggested by some G-33 members prior to the Bali conference would be to allow more explicit adjustment of notifications for excessive inflation (Option 2). Another non-paper by a sub-set of G-33 countries focused on the possibilities for revising the calculation of the MPS component in the AMS through modification of some of the four variables that enter into the MPS calculation subject to the *de minimis* threshold, namely, the *de minimis* level (Options 3a), the external reference price (Option 3b) and eligible production (Option 3c) (BRIDGES WEEKLY, 2013).[4] The idea that administered prices at safety net levels that are below domestic market prices, in the context of procurement of public

---

4 The fourth variable that enters into the MPS calculation is the level of the administered price. Lowering the administered price in order to reduce the MPS element of the AMS

stocks, might be exempt from inclusion in the product-specific AMS (put forward by DIAZ-BONILLA, 2013) is also examined (Option 4). The pros and cons of these alternatives are now briefly discussed.

Option 1 - Make the interim solution permanent

One option for the permanent solution would be to simply make the interim solution permanent but open it to all developing countries. In other words, developing countries which procured food at administered prices for public stock-holding schemes for food security purposes would not be challenged through the dispute settlement mechanism for a breach of their AMS limits due to this expenditure provided they complied with the conditions set out in the Decision. For those countries which had such programmes in place in December 2013, this would be the effect of the Ministerial Decision if no permanent solution can be agreed. This option would extend this protection to all developing country members.

The effect of this solution would be to classify public expenditure on price support as green box expenditure (exempt from AMS limits) and therein lies the major objection. Annex 2 of the AoA which sets out the basis for exemption from reduction commitments states that eligible domestic support measures shall not have the effect of providing price support to producers. Legitimising price support, even in the context of procurement for public stock-holding schemes, would thus be a major breach with one of the main principles behind the definition of exempt support. The opposition to the G-33 proposal in the run-up to the Bali Ministerial shows that it would be difficult to get support for such a major revision of the Agreement. Other options which might give additional flexibility to developing country members without directly breaching this green box requirement should therefore be explored.

Option 2 - Adjusting for inflation

Developing countries are often prone to high rates of domestic inflation which they argue puts them at a disadvantage when calculating their Current Total AMS using the method set out in the AoA. It requires administered prices to be compared to a fixed 1986-88 external reference price in order to calculate the level of AMS support. The argument is that an increase in administered prices which simply reflects domestic inflation has no economic significance in terms of increasing distortions on world markets. In other words, the increase in the administered price does not correspond to an increase in economic price support in real

---

would not address the desire of the proponents of the proposal to gain additional policy space so it is not included as an option.

terms. However, it has a legal impact because, as the administered price is raised to follow the rate of domestic inflation, so the gap with the fixed external reference price becomes larger, increasing the size of the measured current AMS. The non-paper by some G-33 countries would allow WTO members to take into account excessive rates of inflation, for example, higher than 4 percent, in calculating the contribution of food stockholding programmes towards overall farm subsidy commitments (BELLMAN, et al., 2013). This suggestion claims to draw on Article 18.4 of the AoA which reads: "In the review process [of the implementation of Uruguay Round commitments undertaken by the Committee on Agriculture] Members shall give due consideration to the influence of excessive rates of inflation on the ability of any Member to abide by its domestic support commitments."

Inflation is already factored into the AMS calculation, but only partially and depending on how a country has presented its schedule of commitments (WTO, 2000b). In the case of non-exempt budgetary outlays (such as expenditure on a product-specific fertiliser subsidy) where the subsidy payment is indexed to the movement in the inflation index, there is an automatic inflation adjustment in those cases where the subsidy expenditure is below the *de minimis* limit. This is because the value of domestic production (and hence the *de minimis* limit) also rises with the inflation rate. However, this adjustment only partially works in the case of market price support below the *de minimis* limit. Because of the use of the fixed external reference price in local currency, market support in the context of high inflation increases much more rapidly than the nominal value of the *de minimis* limit. Even if a country was compliant in the base year, sooner or later it would breach its commitment even if the only thing that had changed was the indexation of prices to inflation. These conclusions apply a fortiori to those countries where the discipline is specified as a Bound Total AMS because this nominal value remains fixed and does not even increase in line with inflation.

Setting AMS disciplines in nominal terms in the AoA was not accidental and presumably reflected the drafters' intention that over time the value of AMS entitlements would be gradually eroded by inflation. However, if there are significant differences in inflation rates, and some countries experience much higher rates of inflation than others, then the real value of the AMS entitlements of the rapid-inflation countries is eroded much more quickly. One way a country could have protected itself against excessive inflation was to express its Bound Total AMS in terms of a more stable foreign currency by denominating its foreign external reference price in the foreign currency (WTO, 2000b). The country in calculating its Current Total AMS converts its applied administrative prices (in domestic currency) into the foreign currency using the exchange rate of the notification year. For countries with excessive inflation, the rate of domestic price increase and the rate of exchange rate depreciation are usually inversely related. If they happen to exactly offset each other this would provide an automatic inflation adjustment. However, only a handful of developing countries have specified fixed external reference prices or AMS ceilings in foreign currency in their commitments (Argentina, Brazil, Columbia, Costa Rica, Turkey and Venezuela) (WTO, 2000b). Only these countries arguably have access to a built-in inflation adjustment mechanism.[5] Some WTO members have switched from calculating their support in domestic currency in the base period to using a foreign currency in their yearly notifications. For example, by India's switch to notifying in the US currency, its nominal increases in subsidy expenditures and nominal support prices do not show up as increases in notified support to the extent that they are offset by nominal rupee devaluation (GOPINATH, 2011). However, this practice appears to be contrary to the requirement that the AMS should be calculated taking into account the constituent data and methodology used by the country for its base period.

The recourse to Article 18.4 is not sufficient as this Article does not extend any right to a member to modify its domestic support calculations. The paragraph simply allows the Committee on Agriculture to give "due consideration" to the influence of excessive rates of inflation on the ability of any member to abide by its domestic support commitments. The Committee does not have the power either to compel an errant member to bring itself into compliance, or to give protection to a member which finds itself in violation of its commitments. However, some legal weight must presumably be given to the fact that the AoA drafters included this paragraph in the Agreement and thus indicated their awareness that excessive inflation could create problems for a country to remain within its commitments. The suggestion by some G-33 countries would give the same protection to countries which specified their schedule commitments in local currency as is now available to countries which specified in a foreign currency. As this right does not exist in the current AoA, it would require an interpretation of or an amendment to the Agreement specifying how this might be done.

Option 3A - Raise the de minimis level

Another option put forward by some G-33 countries is to raise the limit for developing countries' *de minimis* support currently set at 10 percent of the value of production for all developing countries apart from China, which has a lower limit of 8.5 percent. Their proposal is that this might be raised to 15 percent for all developing countries. Similarly to changing the base period for the fixed external

---

5   The AoA requires that an AMS be calculated by taking into account the constituent data and methodology of the calculations in the country's base period (Article I (a) (ii)).

reference price to a more recent period, this would allow a once-off increase in the ceiling for trade-distorting support for developing countries. However, the distribution of the additional policy space across countries, commodities and programmes would be very different. For example, raising the *de minimis* limit would permit both greater market price support as well as higher expenditure on non-exempt programmes such as input subsidies which do not qualify under Article 6.2, while changing the period for the external reference price only gives greater scope for market price support. It would no doubt be difficult to get political agreement to increase the scope for trade-distorting support in this way. The Doha Round draft modalities (WTO, 2008) provide that the *de minimis* limits for developing countries with Final Bound AMS commitments would be reduced by one-third. However, no reduction in de minimis limits would be required for countries with no Final Bound AMS commitments, or for countries with such commitments which allocate almost all of their support to subsistence and resource-poor producers, or are Net Food-Importing Countries in WTO terms. If a higher *de minimis* limit were pursued in the negotiations on a permanent solution, one possibility might be to allow a temporary increase in the *de minimis* limit but then to gradually reduce it again over time.

Option 3B - Review the basis for the external reference price

A further option could be to review the 1986-88 reference prices that are used as a benchmark in calculating countries' MPS. The G-33 countries say that because this yardstick does not capture increases in food prices over the last few decades, it "grossly exaggerates and overstates the economic subsidy provided." Various alternatives might be envisaged. One option is to change to a variable external reference price. Some G-33 countries proposed that developing countries could use a three-year rolling average of world market prices in their MPS calculation. Alternatively, they might be allowed to use the previous year's average price in the three largest suppliers of foodstuffs in the domestic market. Using a variable external price arguably better captures the economic significance of a country's domestic support policy. For example, it is the benchmark used by the OECD in its calculation of a country's agricultural support (the Producer Support Estimate) (OECD, 2010).

The drawback of using a variable external price is that a country's measured AMS is no longer completely the result of its own policy setting.[6] It also makes it

---

6 The panel ruling on the Brazil-US cotton dispute explains the reasons for the choice of a fixed external reference price for a country's AMS as follows: "... a prime consideration of the drafters was to ensure that Members had some means of ensuring compliance with their commitments despite factors beyond their control" (WTO, 2004b, p. 134).

more difficult for other countries to evaluate the significance of their trading partners' commitments to bind domestic support. World market prices are currently high, making the use of recent prices attractive to those countries which wish to increase their policy space to be able to provide more price support, but this could change in the future. If world market prices start to trend downwards, this could force countries to lower administered prices in order to stay within their AMS limits. For the proponents of variable external reference prices, this is precisely their advantage. Changing the basis for the MPS calculation in this way would de facto lead to a renegotiation of countries' domestic support commitments under the AoA. This makes it less likely that agreement would be reached on a stand-alone basis as part of a permanent solution to the question of public stock-holding for food security purposes, outside of an overall Doha Round agreement.[7] Further, if it were decided to change the legal basis for the MPS calculation in this way, it would be hard to argue that such a fundamental change should be confined to developing countries. Developing countries would then need to evaluate the overall effect of providing greater policy space both for themselves and for developed countries, which have been the greatest users of domestic support to date. An alternative, less dramatic alternative, would be to allow a once-off change in the base period for the fixed external reference price, from 1986-88 to a more recent three-year period during which world market prices were higher. Such a change could be confined to developing countries under special and differential treatment provisions. It is, however, an arbitrary and ad-hoc 'solution' and does not address the fundamental compatibility of WTO rules with food security objectives.

Option 3C - Change the definition of eligible production

One of the parameters in the calculation of the MPS element of the AMS is the quantity of eligible production – which is multiplied by the difference between the fixed external reference price and the administered price to give the value of market price support provided. Countries in their notifications sometimes use total production as the eligible quantity, and sometimes the actual quantity purchased by the government at the administered price. In *Korea-Beef*, the panel argued that, in general, market price support benefits all production of the type and quality supported by the administered price unless there is a legislatively predetermined limit on the quantity eligible for support. In this case, the Appellate Body clarified that it is the quantity that the government has announced is eligible for purchase

---

7 While this objection applies to any proposal to change the basis for the MPS calculation, altering the choice of external reference price would have the most dramatic effect.

which constitutes the eligible production in this case, even if it then actually purchases only a smaller quantity. As here the legal view is consistent with economic reasoning, there does not appear to be a strong case to allow actual purchases as the eligible quantity.

Option 4 - Qualifying the administered price

A different way to address the G-33 concern about policy space has been suggested by DIAZ-BONILLA (2013). His proposal is to clarify the relationship between 'market prices' and 'administered prices' by affirming that administered prices that are at or below market prices should not be seen as providing price "support". The specific language he offered is as follows:

> "Administered prices in the context of this paragraph will be considered rebuttable presumed in compliance with the conditions that they do not offer price support, and therefore, they will not have to be counted against the aggregate measure of support, if they do not exceed the appropriate domestic market price or the import parity equivalent based on the world market price of the product considered."

Such a paragraph would mean that, in order to challenge a food security program of another WTO member, a complainant would have to show that the administered prices were above the appropriate market price, thus offering "price support" to producers. Given volatility in market prices, it is possible that a pre-announced administered price could turn out, in a year with excellent domestic harvests or very low world market prices, to be higher than the subsequent market price. Thus it would be desirable to define the market price on the basis of a rolling average to even out the effects of year-to-year volatility. In any event, as DIAZ-BONILLA (2013) notes, if a developing country is buying significantly above market prices and selling below market prices in order to help poor and vulnerable populations, there is a strong probability it will get into fiscal problems long before a trade case is brought against it.

The justification for this suggestion is that, in developing countries, in the absence of futures markets, the only coordinating device for developing country farmers' expectations about market conditions and for their production decisions may be pre-announced government prices (DIAZ-BONILLA, 2013). Provided these prices are below market prices, then in an economic sense there is no price support.[8] How market prices are defined is clearly relevant. There is an important

---

8 In the presence of uncertainty the existence of a pre-announced support price, by reducing risk, will stimulate supply. In countries with more developed market infrastructure, farmers can shift this risk using futures markets for a small fee. It might be argued that the supply effect of this reduced risk, on its own, is sufficiently small not to violate the green box criteria. However, in Korea-Beef, the panel noted that "the minimum price support will be

difference between the import parity price and the domestic market price which also depends on the extent of applied border protection. The argument that no support is provided by an administered price strictly holds only if it is below the import parity price; even if it were below the domestic market price, if it is above the import parity price then it does provide price support. However, it can be argued that this is not additional to that provided by the border protection alone so that its incremental trade-distorting effect is minimal.[9] Whichever benchmark might be adopted, the general requirement in Annex 2 that measures in the green box cannot provide price support would need to be qualified to allow this exception. The blue box provides a possible precedent in that this shelters an otherwise non-exempt payment to producers by attaching specific conditions which limit its production and trade-distorting impact.

One attraction of this proposal as part of the permanent solution is that it would be straightforward to apply it to public stockholding policies without necessarily requiring a wholesale change in the AMS calculation (though it would be consistent also to apply this interpretation of administered prices to domestic food aid programmes under paragraph 4 of Annex 2). It would allow any developing country to procure foodstuffs from farmers at administered prices provided that (a) it is an integral part of a food security programme identified in national legislation, such that the volume and accumulation of public stocks corresponds to predetermined targets related solely to food security, and (b) provided that the administered price is set below either the import parity or domestic market price. There would, of course, be strong political economy pressures from farm groups over time to raise the level of administered prices, or not to reduce it when import parity or domestic prices are falling. The EU experience of intervention price support is instructive in this regard, where over time the intervention price began to determine the market price and government purchases became a significant element of market demand. The Indian experience is also relevant where purchases of rice and wheat at minimum support prices have resulted in actual stocks that are far in excess of minimum food security norms (HODA and GULATI, 2013). The danger in permitting any kind of economic price support is that there is a

---

    available to all marketable production of the type and quality to which the administered price support programme relates, including where actual market prices are above the administered minimum price level" (para. 827).

9 Thus, a country which procures supplies for food security stocks at domestic market prices which are above import parity prices because of applied border protection provides economic price support to its producers, although such purchases are not constrained in the AoA. It appears that, in India, minimum support prices for rice and wheat have generally been below import parity prices in the past decade (HODA and GULATI, 2013).

strong probability that it leads to trade distortion for the political economy reasons mentioned above.

Conclusion

The compatibility of WTO rules with the food security objectives of developing countries has been a recurring source of controversy. It has now been firmly placed on the WTO agenda by the Decision at the Bali Ministerial Conference to initiate a work programme to find a permanent solution to the question of procurement at official prices for public stock-holding for food security purposes. Although the context for this Decision was a request to exempt such procurement from a product's AMS, the issue raises broader questions about whether and how developing countries should have more policy space to adopt non-exempt and potentially trade-distorting domestic policies. The debate about policy space for developing countries cuts across all three pillars of the AoA. In the market access pillar, it revolves around the role of Special Products and the Special Safeguard Mechanism. In the export competition pillar, it focuses on disciplines on the use of export restrictions. This chapter has focused on the debate on policy space in the domestic support pillar which was the issue addressed in the Bali Ministerial Conference. To what extent, and in what ways, should the domestic support rules and commitments in the AoA be modified to ensure that developing countries are not unduly constrained in pursuing the important goal of food security?

This chapter has reviewed the various proposals made so far for a permanent solution to the treatment of the procurement of public food security stocks at administered prices under WTO rules. Their intention is to provide more flexibility (policy space) for developing countries to pursue currently non-exempt policies where this is justified for food security purposes. Policy space for a WTO member is defined by its right to exempt support under some policies when calculating its current AMS as well as by the size of its limits on AMS support. A country's policy space can be increased either by enlarging the scope of exempt policies, or by increasing the limits on its AMS support.

Three arguments are used to justify increases in developing countries' AMS entitlements. First, any country that uses market price support is disadvantaged in the context of rising food prices by the use of the fixed 1986-88 external reference price as compared to countries that use budget payments to provide farmers with an equivalent level of economic support. This tends to disadvantage those countries, often developing countries, which have few budgetary resources. Second, in spite of the more generous exemptions that developing countries can use in calculating their AMS support, and the higher *de minimis* limits and thresholds, the current distribution of "rights" to use trade-distorting support as between developed and developing countries is arbitrary and without a legitimate justification. WTO ceilings that allow a country a greater right to use trade-distorting support because it was a bigger sinner in the past understandably contributes to the sense of grievance among developing countries that the rules are skewed against them. Third, there is the normative argument that, if particular policies (and especially producer price support polices) are required to ensure the food security of poor populations, then from a right to food perspective these policies should not be constrained by trade rules which are only intended as a means to an end (DE SCHUTTER, 2011).

Countering these arguments is that the deliberate intention of WTO rules is to encourage countries to use less trade-distorting policies in the pursuit of their agricultural and food security policies. With the growing importance of south-south trade, allowing greater scope for developing countries to implement trade-distorting policies will increasingly be to the detriment of other developing countries and their food security. The rapid growth in the amount of domestic support provided by some of the more advanced developing countries underlines the importance of this concern. Other critics point out that price support is an inappropriate way to address food security concerns in the first place. Price support has no impact on subsistence producers and low-income producers are often net purchasers of supported foodstuffs yet these are the groups most vulnerable to food insecurity. Resources devoted to price support can be used to support agricultural production in more productive ways. Yet other critics argue that, given the very different situations of different developing members, with most of them having no immediate risk of breaking their commitments, changing the existing rules would be both hasty and disproportionate to address the concerns raised.

The premise of this chapter is that adapting the rules on exempt policies where this can be shown to be justified to enable developing countries to pursue their food security objectives is a much preferable approach to simply increasing the limits on AMS support. The latter is an ad-hoc approach to dealing with potential inconsistencies between WTO rules and food security policies. It gives no guidance and makes no distinction with regard to how countries might use this increased policy space. Even if an argument can be made that the distribution of entitlements to trade-distorting support as between developed and developing countries is unfair, increasing AMS limits for developing countries (rather than reducing AMS limits for developed countries) runs counter to the overall objective to establish a fair and market-oriented agriculture trading system. If WTO rules are framed in such a way not to restrict developing countries from adopting appropriate policies to address their food security needs, then the case for larger AMS entitlements falls away.

Enlarging the scope of exempted general government services that have relevance to food security was one outcome of the Bali Ministerial Conference. The more controversial question was the treatment of administered prices when procuring public stocks for food security purposes. Two of the proposals reviews in this chapter deserve further consideration in this context. The first would make explicit allowance in the AoA for countries to adjust their measured support for excessive rates of inflation. The drafters of the AoA recognised that this could be a problem, but did not provide a solution. The second would make a distinction between the use of administered prices for price support and as a safety net. Farmers in developing countries are as exposed to price risk but have fewer opportunities to manage this than farmers in developed countries. Where administered prices operate as a safety-net rather than the incentive price to which farmers respond, AoA rules could recognise (along the lines of the blue box) that this use of administered prices is not likely to lead to significant trade distortion and should be permitted. A minimal adaptation of the rules would be to allow this interpretation in the context of public food security stocks, as DIAZ-BONILLA (2013) has suggested. WTO rules already exempt a wide range of policies which address food security needs but are more restrictive about features of those policies that have great potential to distort production and trade. These further amendments could sufficiently adapt the domestic support disciplines to address developing countries' remaining concerns about their ability to pursue their food security goals.

List of References

Anania, G. (2013): Agricultural Export Restrictions and the WTO: What Options do Policymakers have for Promoting Food Security? ICTSD Programme on Agricultural Trade and Sustainable Development; Issue Paper No. 50, International Centre for Trade and Sustainable Development, Geneva.

Bellman, C., Hepburn, J., Krivonos, E. and Morrison, J. (2013): G-33 Proposal: Early Agreement on Elements of the Draft Doha Accord to Address Food Security, ICTSD Programme on Agricultural Trade and Sustainable Development; Information Note. International Centre for Sustainable Trade and Development, Geneva.

Bridges Weekly (2013): WTO – Ag Talks Chair Seeks to Reconcile Conflicting Visions for Bali. Bridg. Wkly. Trade News Dig. 17, 2 May.

De Schutter, O. (2009): Mission to the World Trade Organization. United Nations Human Rights Council.

De Schutter, O. (2011): The World Trade Organization and the Post-Global Food Crisis Agenda: Putting Food Security First in the International Trade System. United Nations Special Rapporteur on the Right to Food, Geneva.

Diaz-Bonilla, E. (2013): Some Ideas to Break the Stalemate on Agricultural Issues at Bali. Food Secur. Portal URL http://www.foodsecurityportal.org/some-ideas-break-stalemate-agricultural-issues-bali (accessed 01.02.2014).

Díaz-Bonilla, E. and Ron, J.F. (2010): Food security, price volatility and trade: some reflections for developing countries. International Centre for Sustainable Trade and Development, Geneva.
G-33 (2012): Statement on Public Stock-holding. URL www.onuperu.org/wto/wto/coass_g_33_16_11_12.pdf (accessed 12.20.2013).
Gonzalez, C. (2002): Institutionalizing Inequality: The WTO Agreement on Agriculture, Food Security, and Developing Countries. Columbia J. Environ. Law 27, 433-489.
Gopinath, M. (2011): India. In: Orden, D., Blandford, D. and Josling, T. (eds.), WTO Disciplines on Agricultural Support. Cambridge University Press, Cambridge.
Häberli, C. (2010): Food security and WTO rules, in: Karapinar, B. and Häberli, C. (eds.), Food Crises and the WTO. Cambridge University Press, Cambridge.
Hoda, A. and Gulati, A. (2013): India's Agricultural Trade Policy and Sustainable Development. International Centre for Trade and Sustainable Development, Geneva.
Howse, R. (2013): The Peace Clause a Trojan Horse. Int. Econ. Law Policy Blog. URL http://worldtradelaw.typepad.com/ielpblog/2013/11/the-peace-clause-a-trojan-horse.html (accessed 01.02.2014).
ICSTD (2013): Historic Bali Deal to Spring WTO. Global Economy Ahead, Bridges Daily Update 5, 7 December, URL http://ictsd.org/i/wto/wto-mc9-bali-2013/bridges-daily-updates-bali-2013/180991/ (accessed 01.04.2014).
Matthews, A. (2012a): The Impact of WTO Agricultural Trade Rules on Food Security and Development: An Examination of Proposed Additional Flexibilities for Developing Countries, in: McMahon, J. and Desta, M. (eds.), Research Handbook On The WTO Agriculture Agreement: New and Emerging Issues in International Agricultural Trade Law. Edward Elgar Publishing, London.
Matthews, A. (2012b): Trade Agreements, WTO Rules and Food Security. Presented at the International Agricultural Trade Research Consortium (IATRC) Annual Meeting, December, San Diego.
Matthews, A. (2013): Doha Negotiations on Agriculture and Future of the WTO Multilateral Trade System. Presented at the Challenges for the Global Agricultural Trade Regime after Doha, 135th EAAE Seminar, Belgrade.
OECD (2010): OECD's Producer Support Estimate and Related Indicators of Agricultural Support: Concepts, Calculations, Interpretation and Use (The PSE Manual). Organisation for Economic Cooperation and Development, Paris.
Oryza (2013): Bali Package on Foodgrain Stockpiling Against WTO Ideals. Claim Pakistan Rice Growers, URL http://oryza.com/news/bali-package-foodgrain-stockpiling-against-wto-ideals-claim-pakistan-rice-growers (accessed 01.03.2014).
South Centre (2013): The WTO's Bali Ministerial and Food Security for Developing Countries: Need for Equity and Justice in the Rules on Agricultural Subsidies. South Centre, Geneva.
Wall Street Journal (2013): India Says Developing Countries Want More Fropm WTO. URL http://online.wsj.com/news/articles/SB10001424052702304579404579233640256964118 (accessed 1.15.2014).
Wolfe, R. (2010): Sprinting during a Marathon: Why the WTO ministerial failed in July 2008. J. World Trade 44, 81-126.
WTO (2000a): Agreement on Agriculture – Special and Differential Treetment and a Development Box. Proposal to the June 2000 Special Session of the Committee on Agriiculture by

Cuba, Dominican Republic, Honduras, Pakistan, Haiti, Nicaragua, Kenya, Uganda, Zimbabwe, Sri Lanka and El Salvador, G/AG/NG/W/13, Geneva.

WTO (2000b): Inflation and Exchange Rate Movements in the Context of Domestic Support Commitments, G/AG/NG/S/19. World Trade Organisation, Geneva.

WTO (2004a): Doha Work Programme, Decision Adopted by the General Council on 1 August, 2004, WT/L/579. World Trade Organisation, Geneva.

WTO (2004b): United States - Subsidies on Upland Cotton, Report of the Panel of the Dispute Settlement Body, WT/DS267/R. World Trade Organisation, Geneva.

WTO (2008): Revised Draft Modalities for Agriculture, TN/AG/W/4/Rev.4, 6 December, World Trade Organisation, Geneva.

WTO (2013a): Public Stockholding for Food Security Purposes: Ministerial Decision of 7 December, WT/MIN(13)/38. World Trade Organisation, Geneva.

WTO (2013b): Compliance with Notification Obligations: Note by the Secretariat, G/AG/GEN/86/Rev.15. World Trade Organisation, Geneva.

Production and Wastage of Rice in Bangladesh

Khandaker M. M. RAHMAN, Mohammad I. A. MIA, Mohammad Z. ABEDIN and Mohammad Z. RAHMAN

Abstract

This study has been conducted in Bangladesh in the year 2010 to assess potential, harvest and post-harvest losses of rice that occurred during production, harvest and post-harvest operations like cutting, field drying, transporting from field to threshing yard, threshing, winnowing, drying, in-store, out-store, transporting, marketing etc. The main motivation of this study was to estimate stock availability towards formulating policy options for national food security. Total harvest losses for *Aus, Aman, Boro* and all rice were respectively 6.33%, 6.30%, 7.12% and 6.65% of total gross production whereas total post-harvest losses were 9.57%, 8.90%, 10.16% and 9.52%. Total wastages for *Aus, Aman, Boro* and all rice were respectively 15.90%, 15.20%, 17.28% and 16.17% respectively. The total seed, feed and wastages for the above crops were respectively 21.88%, 19.31%, 19.91% and 19.64%. Total losses of rice including potential loss were the highest for *Boro* (33.91%) followed by *Aman* (30.31%) and *Aus* (25.88%) respectively with the total loss of all rice at 31.64%. Per farm marketable surpluses for *Aus, Aman, Boro* and all rice were respectively about 21%, 52%, 59% and 54% of gross production. The stock availabilities per farm at the end of each crop season were respectively about 8%, 5%, 2% and 4%. A package of policy options and recommendations such as developments of less grain shattering high yielding variety and rural infrastructure, irrigation facilities, timely cutting of rice, threshing, winnowing and drying with mechanised threshers in well protected paved floor etc. were suggested to reduce the total wastage of rice.

Introduction

Bangladesh is one of the best naturally fertile lands where huge quantities of various crops are produced to feed the growing millions of population. Rice is the staple food accounting for about 93 percent of the total food produced, about 70 percent of average calorie intake and 35 percent household expenditure. Rice production is the largest contributor to farm income, while related trade and commerce are one of the largest sources of rural non-farm income (AHMED, 2001). Bangladesh is the fourth largest rice producer in the world (FAO, 2010). In spite of this, the country is languishing with food deficit and each year the country has been importing over one million metric tons of rice at the expense of hard-earned foreign currency (BBS, 2009). The discrepancy between rice production and demand has been widening due to growing population and indiscriminate use of agricultural land for non-agricultural purposes. Although our land is favourable for rice production, we cannot use the full production potentiality for a number of factors. We are losing about 16% of rice production due to technical inefficiency effect (RAHMAN, 2002; RAHMAN et al., 1999). In addition, a considerable amount of rice (paddy) is lost in each stage of production especially in harvest, processing and storage stages. Previous studies showed that the losses of rice in post-harvest operations in Bangladesh were more than 13 percent (CALVERLEY, 1994; QUASEM and SIDDIQUEE, 2009). SAMAJPATI and SHEIKH (1980) found that the food grain loss in rice during processing and storage are as high as 22% of the total production. The losses are higher in Bangladesh compared to other developing countries where better storage systems are developed. If harvest loss is minimised up to 50 percent of the present level by adopting BRRI developed power tiller and power hand reapers resulting a saving of 0.843 million tons of paddy (RAHMAN, 2009). The losses are mainly caused by technical inefficiencies in production, harvest and post-harvest losses and losses due to non-food usages at different sectors like keeping for seed need, industrial usage, feeding livestock, poultry and fisheries. A number of biological, technological, human resources and material input factors have been affecting the production efficiency of rice which is the main crop to be made available as food security for the population of Bangladesh. Moreover, a substantial portion of the produced rice is lost during post-harvest operations including transporting, storage, marketing, etc., to make access to consumers. A portion of rice also goes to non-human food usage.

Previous studies in other developing countries justified our results at different stages of post-harvest operations. FAO (2004) referred post-harvest activities as the suite of processes "from the floor to the fork"- threshing, milling, processing, market transport and cooking. Although much progress has been made in the prevention of post-harvest losses in rice, nevertheless rice losses average between 15 and 16 percent in developing countries. These rice losses are significant during critical operations such as drying, storage and milling.

MANDOZA and GUMMERT (2007) in their study "Fighting Asia's post-harvest problems" narrated that the fate of rice after harvest is a crucial but often-neglected part of the production chain. Now, a major effort to overcome post-harvest problems is gaining momentum. They say that if you're a rice farmer anywhere in Asia, you are likely to experience high post-harvest grain losses. Total losses from harvest to market can reach 30-50%, which means that, conservatively, farmers are losing around US$30 per ton of rice harvested. For an average four-member farm family, an additional $30 can go a long way. MANDOZA and GUMMERT (2007) further reported that the studies by the International Rice Research Institute (IRRI) in Cambodia, Indonesia, and the Philippines have found that post-harvest losses occurred mainly because of spoilage and wastage at the farm level, delay in drying, poor storage, poorly maintained or outdated rice mills, and losses to pests throughout the post-harvest chain. These losses result in lower quality rice for consumption or sale, smaller returns to farmers, higher prices for consumers, and greater pressure on the environment as farmers try to compensate by growing more rice.

A study of IRRI in the Philippines carried out by MEJIA et al. (NAPHIRE, 1997) showed that only harvest loss by traditional hand cutting method at optimum maturity stage for different rice varieties ranged from 6% to 17%. This loss further increased substantially with delayed cutting of paddy. A study by FAO (2004) revealed that only harvest losses by traditional hand cutting in Thailand and Myanmar were respectively 9.3% and 1.9% respectively.

CALVERLEY (1994) evaluated post- harvest losses in several countries of Central and South-Eastern Asia comprising eleven FAO projects. He observed that average loss of harvest, threshing, drying, storage and milling loss was 13.56% by simple arithmetic calculation and 12.89% by cumulative addition. Another study of FAO Corporate Document Repository (2004) in China (ZHEJIENG, IDRC study survey) covering three rice seasons (1987-1989) observed that average losses of six operations viz., threshing, drying and cleaning, storage, transport and milling was 14.81% of total production, whereas, the in-store loss in Bangladesh was as high as 19.7%, out of which 12-13% loss was caused by insects and rodents only. These loss figures were confirmed by a World Bank study on commercial paddy storage in Bangladesh as reported by FAO Corporate Document Repository (2004).

There is another type of loss in the production of rice which occurs in the production process due to lack of the realisation of full potential of productive inputs. This loss is called potential production loss, which occurs mainly due to technical inefficiency effects. Farmers can save resources by using optimum combination of inputs to attain maximum possible production from these given set of inputs. RAHMAN et al. (1999) studied technical efficiency of rice farmers in Bangladesh using the stochastic frontier production function methodology. They observed that the technical efficiencies for the *Aus*, *Aman* and *Boro* rice were respectively 93%, 80% and 86% at the aggregate level. That is, farmers could increase production of *Aus*, *Aman* and *Boro* rice by 7%, 20% and 14% respectively using the same inputs and available resources. DEB and HOSSAIN (1995) estimated technical efficiency of rice growers in Bangladesh using the same stochastic frontier. They observed that average technical efficiency was 84% at the aggregate level.

The farmers generally store paddy/rice mostly using traditional storage structures. Farmers' level storage structures are not perfect and stored rice is subjected to damage due to many biotic and abiotic factors. The net availability of rice is considerably less than gross production due to these factors. A comprehensive research was, therefore, needed to achieve field level information on production and losses at stages of harvest and post-harvest operations, storage including other food and non-food usage that would contribute to formulate appropriate policy guidelines in ensuring food availability from farm level production of rice. This was based on sample survey covering fourteen intensive rice growing districts across the country.

Objectives

The overall objective of the study was to assess potential, harvest and post-harvest losses that occurred during production, processing and storage operations towards estimating stock availability and thus formulating policy options helpful for national food security. The study was undertaken to attain the rice and farm specific following objectives:

- To assess technical efficiency in the production of rice;
- To assess potential losses due to technical inefficiency in producing rice;
- To assess losses of rice at different stages of harvest and post-harvest operations;
- To assess non-food usages of rice;
- To assess losses in transporting and marketing;
- To assess present stock availability;
- To formulate policy guide lines for improving efficiency of rice growing farmers for realizing full potential of production and reducing harvest, post-harvest losses and non-food usage in addressing food availability in the country.

Methodology

In this study, we were mainly interested to know the rice stock situation in Bangladesh. In the production a portion of rice (paddy) is lost at harvest and post-harvest operations by numerous factors, actors, storage up to the harvest of rice in the next session. A portion of rice grain is used for feeding poultry and livestock, and donating to various religious or social organizations, institutions, beggars and poor people in the locality. A portion of rice is wasted too in transportation, marketing and consumption processing. The researchers estimated technical efficiency, losses and usages for ascertaining total rice stock availability at individual farm level. All these dimensions of study were classified under the following major headings:

- Technical efficiency;
- Harvest and post-harvest rice losses and usages;
- In-store losses and losses in transportation and marketing;

- Farm stock availability.

Moreover, in order to address major issues in formulating policy we analysed the grass root level stakeholders' suggestion for increased stock availability towards increased marketable surplus for consumer. In order to reach the objectives we followed the steps as described below.

Data and study area

For the study we selected 96 villages under 26 sub-districts of 14 civil districts across all the divisions of the country representing more than 14 thousand rice farm households of marginal, small, medium and large categories. Selection of the areas was purposive based on the major and intensive production, accessibility, cropping patterns and surplus of rice in the districts as the districts contributed to about two-thirds of the country's rice production (BBS, 2009). The sample households were selected randomly using stratified random sampling technique with arbitrary allocation. The study covered 5 regions. Region 1 consisted of Dinajpur, Rangpur and Bogra districts. Region 2 covered Rajshahi, Natore and Naogaon districts. Region 3 included Mymensingh, Sherpur and Kishoregonj districts. Region 4 covered the districts of Sylhet, Camilla, Chittagong and Region 5 covered Jessore and Barisal districts.

The overall sample household was 2.27% of the total population of households. The farm households were classified into four groups, such as Marginal farmers having land holdings ranging from 0 to less than 0.4 hectare, Small farmers having land holdings ranging from 0.4 to 0.99 hectare, Medium farmers having land holdings ranging from 1 to 3 hectares, and Large farmers having land holdings more than 3 hectares. The data were collected from the separate *Aus, Aman* and *Boro* rice households in 2010. The total sample size was 1360. For collection of data we used separate pretested questionnaires and PRA check lists.

Analytical technique

For analysing the data both descriptive methods and functional models were used. The descriptive methods were mostly used for analysing data of assessing losses incurred in harvest, post- harvest, and storage and out storage operations including marketing and consumption processing (milling) operations. The functional models were Cobb-Douglas stochastic frontier production function and technical inefficiency effect model. The assessed parameters in the study are described below:

- Potential production (output) = maximum attainable output (MAO) with available inputs and existing technology used in the production of *Aus, Aman* and *Boro* rice by individual farmers.
- Observed output = output received for a particular type of rice, *Aus* or *Aman* or *Boro* produced by the individual farmers.
- Technical efficiency (TE) = produced output of rice / MAO.
- Technical inefficiency = 1 – TE.
- Total seed-feed-wastage = seed need + feed + wastage.
- Feed = usages for feeding: cattle + goat + poultry + pet animals etc.
- Wastage = harvest loss + post-harvest loss.
- Harvest loss = loss due to: cutting + field drying + bundling + transporting from field to farm/threshing yard.
- Post-harvest loss = loss due to: threshing + winnowing/cleaning + drying + bulk handling for storage + in-store + out-store + transporting and marketing.
- In-store loss = loss due to biotic and abiotic factors.
- Out-store loss = loss due to handling + transporting.
- Donations = donation to: institution + organizations + persons + relative.
- Marketable surplus = net production – total household use.
- Net production = gross production – total seed-feed-wastage.
- Total household use = consumption requirements + donations.
- Stock availability of rice = marketable surplus – sale + purchase.

Results and Discussion

Productivity and profitability of producing rice

A summary statistics on some farm-specific or socioeconomic variables used in stochastic frontier and inefficiency effect model is presented in Table 1. All variables are expressed as per farm basis. Table 1 reveals that mainly medium aged farmers were engaged in farming practices and average age of farmers was 46.45 years with significant variations among crops ($F = 5.51^{**}$). Education levels of farm operators insignificantly ($F = 1.03$) varied among crops and the average education level was 7.01 years of schooling. The average experience of farming was 25.24 years. The distribution of cultivable land under rice production was quite dissimilar among crops ($F = 21.50^{**}$). The farmers diverted the largest area for *Boro* rice cultivation (239.34 decimals) with an average rice area at 220.74 decimals for all rice crops. Similarly, total land under households was also the highest (327.13 decimals) for the farmers producing *Boro* rice (Table 1).

Tab. 1:   Rice type wise distribution of different socio-economic variables

Type	Sample size	Age (year)	Education (year of schooling)	Experience (year)	Area under Production (decimal)	Total land under household (decimal)
				Mean		
Aus	209	44.02 (12.84)	6.87 (4.12)	23.80 (11.95)	122.14 (106.78)	259.73 (180.67)
Aman	587	46.55 (12.09)	6.86 (3.99)	25.25 (12.43)	237.98 (247.50)	325.88 (290.63)
Boro	564	47.26 (11.72)	7.22 (5.26)	25.76 (12.81)	239.34 (258.19)	327.13 (286.77)
Total	1360	46.45 (12.10)	7.01 (4.58)	25.24 (12.52)	220.74 (239.88)	316.24 (275.80)
F-value		5.51**	1.03	1.85	21.50**	5.22**

*Figures in the parentheses indicate standard deviations. **indicates significance at 0.01 probability level*

Source: Own calculation

Table 2 summarises the farm inputs used for the production of rice crops per hectare basis among different crops. Some of the inputs are expressed in money values. The farmers in *Boro* rice used relatively higher amount of labour (164.27 man-days) per hectare followed by farmers in *Aman* and *Aus* rice, respectively (F = 0.293). Farmers in *Aman* crop used relatively higher amount of seed (48.87 kg) followed by farmers in *Boro* and *Aus* crop respectively. Farmers in different crops used different amounts of fertiliser with significant variations (F = 120.75**). Farmers in *Boro* crop used the highest amount of fertiliser (414.53 kg) whereas farmers at the aggregate level used 350.18 kg of fertiliser. Manure uses were also found to be significantly different in different rice crops (F = 6.83**). The average ploughing cost was Tk.4189.97. There were significant differences (F = 561.82**) in irrigation cost among the three types of rice, with the highest cost (Tk.4440.34) in *Boro* rice. Insecticide cost also showed significant differences (F = 94.77**) with highest cost in *Boro* crop (Tk.1140.67) followed by *Aus* (Tk .924.54) and *Aman* (Tk.910.87) respectively with aggregate cost at Tk.1008.27.

Tab. 2: Rice type wise per hectare uses of different farm inputs

Rice type	Sample Size	Labour (Man-days)	Seed (kg)	Fertilizer (kg)	Manure (kg)	Ploughing cost (Tk.)	Irrigation cost (Tk.)	Insecticide cost (Tk.)
Aus	209	160.65 (58.86)	47.89 (15.34)	311.02 (124.15)	3732.58 (2808.25)	4155.62 (1957.76)	1668.43 (2161.08)	924.54 (307.57)
Aman	587	163.12 (59.33)	48.87 (16.32)	302.29 (136.44)	3014.52 (2560.68)	4248.80 (1876.55)	1115.37 (1765.14)	910.87 (288.08)
Boro	564	164.27 (57.43)	48.16 (19.16)	414.53 (121.95)	3414.64 (2626.03)	4141.47 (2015.89)	4440.34 (1514.03)	1140.67 (306.32)
Total	1360	163.22 (58.45)	48.43 (17.41)	350.18 (139.61)	3290.80 (2638.30)	4189.97 (1947.12)	2579.25 (2344.91)	1008.27 (318.73)
F-value		0.293	0.359	120.75**	6.83**	0.475	561.82**	94.77**

Figures in the parentheses indicate standard deviations. ** *indicates significance at 0.01 probability level*
Source: Own calculation

Table 3 summarises per hectare basis labour and rice production costs, gross return, net return and benefit cost ratio (BCR) among different types of rice. The farmers in *Boro* rice production paid significantly higher amount of labour cost (Tk.26404.33) per hectare (F = 3.03*). Similarly, farmers in *Boro* crop incurred the highest amount of production cost (Tk.64836.39) among all crops (F=39.25**). *Boro* crop exhibited the highest production per hectare (5928.09 kg) followed by *Aman* (3886.41 kg) and *Aus* crop (3291.36 kg) respectively. Per hectare gross return was significantly (F = 249.38**) the highest for *Boro* rice (Tk 88477.77). Similarly, the highest amount of net return was found in *Boro* crop (Tk. 23641.38) followed by *Aman* (Tk.7786.04) and *Aus* rice (Tk.1082.44), whereas the overall net return was Tk.13331.16. Consequently, benefit cost ratio was significantly (F = 39.83**) the highest for *Boro* rice (1.47).

Technical efficiency of producing rice

Table 4 shows frequency distribution of farm-specific technical efficiency estimates for *Aus, Aman* and *Boro* rice from Cobb-Douglas stochastic frontiers. The results revealed that about 100% sample farmers of *Aus* rice were obtaining outputs which were very close to the maximum output estimated through frontier (efficiency is 95% to 100%) and there were about 24% of *Aman* rice farmers whose technical efficiency levels ranged from 95% to100%, whereas only about

1% farmers produced *Boro* output at 95-100% efficiency level. For *Aus* rice technical efficiency varied from 95% to 100% and for *Aman* rice it varied from 70% to 100%, whereas for *Boro* rice it varied from 50% to 100%.

Tab. 3: *Rice type wise per hectare cost, return and benefit cost ratio (BCR)*

Crop	Per hectare cost and return (Tk.)					Benefit-Cost Ratio (BCR)
	labour cost	production	gross return	cost	net return	
Aus	23993.24 (11842.50)	3291.36 (847.91)	57917.81 (16816.99)	56835.37 (16408.04)	1082.44 (24868.82)	1.11 (0.48)
Aman	25384.48 (12540.98)	3886.41 (841.38)	64788.16 (15779.87)	57002.13 (16516.63)	7786.04 (21275.57)	1.24 (0.53)
Boro	26404.33 (12471.01)	5928.09 (1413.93)	88477.77 (26613.87)	64836.39 (15677.75)	23641.38 (34136.43)	1.47 (0.64)
All	25593.62 (12425.72)	4641.66 (1567.51)	73556.57 (24643.60)	60225.42 (16605.44)	13331.16 (29207.25)	1.32 (0.58)
F-value	3.03*	662.17**	249.38**	39.25**	70.67**	39.83**

Figures in the parentheses indicate standard deviation. ** *and* * *indicate significant at 0.01 and 0.05 probability level, respectively.*

Source: Own calculation

The average technical efficiency coefficients for *Aus, Aman, Boro* and all rice were respectively 0.96, 0.89, 0.86 and 0.88. The maximum efficiency coefficients attained for *Aus, Aman, Boro* and all rice crops were respectively 0.97, 0.98, 0.96 and 0.98, whereas the minimum efficiency scores for the above crops were respectively 0.95, 0.73, 0.53 and 0.53 (Table 5). RAHMAN et al. (1999) found similar results while studying technical efficiency of rice farmers in Bangladesh using the same methodology. They observed that the technical efficiencies for the *Aus, Aman* and *Boro* rice were respectively 93%, 80% and 86% at the aggregate level. DEB and HOSSAIN (1995) estimated technical efficiency of rice growers in Bangladesh using stochastic frontier. They observed that average technical efficiency was 84% at the aggregate level. BANIK (1994 found that the technical efficiency in irrigated modern variety of *Boro* rice cultivation was 0.82.

Assessing losses of rice at stages of harvest and post-harvest operations

Assessing losses of *Aus, Aman* and *Boro* rice was used to indicate losses incurred during different stages of harvest and post-harvest operations for different rice type in different cropping seasons. Losses of rice at harvest operations was used to indicate the losses incurred during cutting paddy, drying cut rice in the field for some days and amount lost for carrying from farm to yard and amount lost for bundling and amount destroyed by animal, insects were measured.

Tab. 4: *Frequency distribution of crop specific technical efficiency estimates from Cobb-Douglas stochastic frontiers*

Efficiency Level (%)	Rice type (Figurs in the parentheses indicate percentages)		
	Aus	Aman	Boro
50-55	-	-	1 (0.2)
55-60	-	-	5 (0.9)
60-65	-	-	5 (0.9)
65-70	-	-	18 (3.2)
70-75	-	2 (0.3)	16 (2.8)
75-80	-	18 (3.1)	42 (7.4)
80-85	-	63 (10.7)	87 (15.4)
85-90	-	131 (22.3)	186 (33)
90-95	-	232 (39.5)	199 (35.3)
95-100	209 (100)	141 (24)	5 (0.9)
Total number of farms	209 (100)	587 (100)	564 (100)

Source: Own calculation

Tab. 5: Rice type wise technical efficiency coefficients

Efficiency parameter	Rice type			
	Aus	*Aman*	*Boro*	All
Maximum	0.97	0.98	0.96	0.98
Minimum	0.95	0.73	0.53	0.53
Mean	0.96	0.89	0.86	0.88

Source: Own calculation

Rice type and farm category wise average harvest loss

Rice type and farm category wise average harvest loss has been estimated (Table 6). The percentages of average losses of all types of rice and farms have been estimated by converting the amount of losses per 40 kg production. The loss of *Boro* rice was the highest in all farm categories. However, the average loss of all the types of rice was the highest in large farms and the lowest in marginal farms. Average losses of *Aus, Aman* and *Boro* and all rice were 5%, 5.05%, 5.58% and 5.26% respectively.

Tab. 6: Rice type and farm category wise average harvest loss

Farm category	Harvest loss (%)			
	Aus	*Aman*	*Boro*	Average
Large	4.96 (3.18)	5.53 (3.74)	5.67 (4.25)	5.51 (3.88)
Medium	4.79 (2.78)	5.01 (3.40)	5.72 (4.52)	5.27 (3.89)
Small	5.11 (2.87)	5.23 (3.29)	5.47 (3.67)	5.31 (3.39)
Marginal	5.12 (3.33)	4.69 (3.19)	5.52 (3.88)	5.10 (3.52)
Average	5.00 (3.00)	5.05 (3.27)	5.58 (3.88)	5.26 (3.50)

Figures in the parentheses indicate standard deviations
Source: Own calculation

Rice type and farm category wise transportation loss of harvested rice from field to threshing yard

Since harvested rice was carried from field to threshing yard/farm yard by traditional methods, a significant amount of rice was lost during this operation. Transporting rice from field to threshing yard for further operations were done mainly by means of human and animal power and sometimes by low horse powered vehicles, like power tiller van, small tractor and truck. Harvested rice was mainly transported by head load, shoulder carrying using bamboo poles, carts poles by animal or man, boats, rickshaw vans where the harvested stalks were bundled or tied or piled. The study revealed that total losses in this stage were 1.33% in *Aus*, 1.25% in *Aman* and 1.54% in *Boro* rice. Carrying of rice stalk from the field for the large farms was mainly done by waged labourers where these activities for the small and marginal farms were done mainly by the farmers themselves. Therefore, they tried their best to keep transportation loss at the least (Table 7).

Tab. 7:   Rice type and farm size wise average transportation loss

Farm category	Transportation loss (%)			
	Aus	Aman	Boro	Average
Large	1.53 (1.43)	1.36 (1.11)	1.60 (1.71)	1.48 (1.43)
Medium	1.32 (1.09)	1.35 (1.27)	1.61 (1.69)	1.45 (1.44)
Small	1.36 (1.01)	1.21 (1.20)	1.53 (1.42)	1.37 (1.28)
Marginal	1.24 (1.08)	1.15 (1.05)	1.46 (1.49)	1.29 (1.26)
Average	1.33 (1.09)	1.25 (1.17)	1.54 (1.55)	1.39 (1.34)

Figures in the parentheses indicate standard deviations
Source: Own calculation

Rice type and farm category wise average threshing, winnowing and drying loss

Rice type and farm category wise average threshing, winnowing and drying losses are given in Table 8. For the three types of rice losses were the lowest for marginal farmers (2.5% for *Aman*, 2.8% for *Aus* and 2.99% for *Boro* rice). The average threshing, winnowing and drying loss for *Aman* rice was again the lowest while it was the highest for *Boro* rice (3.06%). However, the average loss of all the three types of rice was 2.89%. In threshing, winnowing and drying losses occurred because some grains remained in the bundle panicles but no repetition of threshing was done. Some grains were eaten by birds and domestic fowls. During drying some grains spilled out of the bags during transportation from the threshing yard to place where threshed rice is stacked for drying for several days' until it reaches minimum moisture content. Some grains are also spilled outside the drying area during drying.

Tab. 8: Rice type and farm category wise average threshing, winnowing and drying loss

Farm category	Threshing, winnowing and drying loss (%)			
	Aus	Aman	Boro	Average
Large	3.15 (1.40)	3.01 (2.07)	2.79 (1.93)	2.94 (1.92)
Medium	3.08 (2.07)	2.83 (2.11)	3.12 (2.15)	2.99 (2.12)
Small	3.01 (1.91)	2.65 (2.02)	3.14 (2.06)	2.9 (2.03)
Marginal	2.80 (1.93)	2.50 (1.96)	2.99 (2.10)	2.75 (2.02)
Average	2.98 (1.92)	2.70 (2.03)	3.06 (2.09)	2.89 (2.05)

Figures in the parentheses indicate standard deviations
Source: Own calculation

Rice type and farm category wise average storing preparatory loss

The loss of marginal farmers of *Aus* rice was the highest (1.02%), the next loss was for *Boro* rice (0.99%) for the same category of farmers (Table 9).

Rice type and farm category wise average in-store loss

Significant losses in quantity and quality in in-stored situation occurred through the activities of microorganisms, insects, mites and rodents. Traditional and inefficient methods of harvest, threshing, drying, storing and processing caused about 13% post-harvest loss of the food grain which is about 3.51 million metric tonnes per year (QUASEM and SIDDIQUEE, 2009). Rice can be stored safely for two to three months at 14% moisture content. For longer storage, the rice should be dried to moisture content of about 12% (BANGLAPEDIA, 2008). ALAM et al. (2007) found that farmers in their storage structures and containers got lost of paddy of 2.33%. Losses occurred during 3.05 months to 7.24 months (average 5.55 months) of in-store situation had been estimated in our study (Table 10). The average in-store loss of *Aus, Aman, Boro* and all rice was respectively 3.68%, 3.80%, 4.12% and 3.92%.

Tab. 9: *Rice type and farm category wise average storing preparatory loss*

Farm category	Average storing preparatory loss (%)			
	Aus	Aman	Boro	Average
Large	0.93	0.62	0.85	0.76
Medium	0.74	0.66	0.97	0.81
Small	0.82	0.67	0.89	0.78
Marginal	1.02	0.63	0.99	0.84
Average	0.86	0.65	0.94	0.80

Source: Own calculation

Tab. 10: *Rice type and farm category wise average in-store loss due to biological factors*

Farm category	In-store loss (%)			
	Aus	Aman	Boro	Average
Large	4.32 (5.55)	4.32 (3.52)	4.70 (4.77)	4.48 (4.35)
Medium	3.9 (2.99)	3.79 (2.42)	4.06 (3.69)	3.92 (3.10)
Small	3.54 (2.12)	3.99 (3.29)	4.16 (3.54)	4 (3.28)
Marginal	3.4 (1.97)	3.36 (3.48)	3.92 (3.63)	3.59 (3.36)
Average	3.68 (2.85)	3.80 (3.15)	4.12 (3.75)	3.92 (3.38)

Figures in the parentheses indicate standard deviations
Source: Own calculation

Rice type and farm category wise average out-store loss

The estimate of out-store loss include de-storing, filling in the bags and sacks, transporting and marketing of rice (Table 11). The loss of *Aus* (2.05%) and *Boro* (2.04%) rice were higher than those of *Aman* rice (1.75%).

Tab. 11: Rice type and farm category wise average out-store loss

Farm category	Average out-store loss (%)			
	Aus	Aman	Boro	Average
Large	2.07	2.05	1.88	1.98
Medium	1.83	1.69	2.23	1.94
Small	2.13	1.79	1.94	1.9
Marginal	2.18	1.63	2.00	1.87
Average	2.05	1.75	2.04	1.91

Source: Own calculation

Rice type and farm category wise donation

The farm households donated a portion of their rice to social, education and religious organisations, institutions like cooperatives and persons like beggar, neighbouring poor and relatives. The comparisons by type and farm category revealed that donations of *Aus* rice were the highest (4.25%) and *Boro* rice were the lowest (3.65%) as shown in Table 12. On an average, 3.93% rice was donated to these organizations, institutions and persons. Donated *Boro* rice was less in percentage but was large in volume.

Tab. 12: Rice type and farm category wise average donation

Farm category	Donation (kg)			
	Aus	Aman	Boro	Average
Large	93.20 (3.63)	266.37 (3.44)	471.38 (3.01)	328.92 (3.28)
Medium	72.04 (3.80)	123.08 (3.18)	170.52 (3.57)	134.95 (3.44)
Small	57.23 (4.28)	86.44 (4.20)	113.56 (4.02)	93.35 (4.14)
Marginal	42.08 (4.91)	59.06 (5.18)	74.03 (3.56)	62.41 (4.47)
Average	60.93 (4.25)	108.3 (4.08)	158.05 (3.65)	121.65 (3.93)

Figures in the parentheses indicate percentages of donation from per farm gross production
Source: Own calculation

Rice type and farm category wise non-human food usages

The farm households also used a portion of total production for feeding domestic animals and birds like cow, goat, buffalo, duck and pigeon. The percentage of rice used in feeding animals and birds are pooled under the heading of non-human food usages (Table 13). The highest percentage (3.90%) of non-human food was made by using *Aus* rice followed by *Aman* rice. On an average 2.58% of rice was used for feeding the domestic animals and birds.

We have found that the percentage of non-human food usages was the highest (3.30%) in marginal farm households but the lowest (1.97%) in large farm households.

Tab. 13: Rice type and farm category wise average non-human food usages

Farm category	Amount of non-human food usages (kg)			
	Aus	Aman	Boro	Average
Large	90.85 (4.97)	186.93 (2.16)	122.99 (0.77)	145.67 (1.97)
Medium	57.1 (3.23)	90.37 (2.21)	81.13 (1.47)	81.31 (2.06)
Small	34.86 (2.62)	56 (2.65)	63.4 (2.69)	55.97 (2.67)
Marginal	44.45 (5.63)	38.41 (3.46)	43.62 (2.22)	41.53 (3.30)
Average	49.89 (3.90)	74.85 (2.68)	69.84 (1.99)	68.94 (2.58)

Figures in the parentheses indicate percentage of non-human food usages
Source: Own calculation

Estimating total wastage (harvest and post-harvest losses)

The estimated harvest losses for *Aus, Aman* and *Boro* rice were 6.33%, 6.30% and 7.12% respectively. Total average harvest loss for all rice was 6.65%. The post-harvest losses for *Aus, Aman* and *Boro* rice were 9.57%, 8.90% and 10.16% respectively. Total average post-harvest loss for all rice was 9.52%. Total wastage was calculated by summing up the losses incurred during harvest and post-harvest operations. The wastages for *Aus, Aman* and *Boro* rice were 15.90%, 15.20% and 17.28% respectively. On an average, the wastage was 16.17% for all types of rice. Seeds kept for *Aus, Aman* and *Boro* rice were 2.49%, 1.85% and 1.24% respectively with the average being 1.58%. The feed usage for cattle, poultry, goat, pet

Production and Wastage of Rice in Bangladesh 201

animals, etc. of *Aus, Aman* and *Boro* rice were 3.49%, 2.26% and 1.39% respectively with the average being 1.89% for all rice type (Table 14). The total seed, feed and wastages (SFWs) were 22% for *Aus* rice, 19% for *Aman* rice and 20% for *Boro* rice.

Tab. 14: Rice type wise average seed, feed and wastage (SFW)

	Items	Average per farm SFW (%)			
		Aus	Aman	Boro	Average
A	Harvest operations (Cutting, field drying & bundling)	5.00 (3.00)	5.05 (3.27)	5.58 (3.88)	5.26 (3.50)
	Transporting from field to farm yard/threshing yard	1.33 (1.09)	1.25 (1.17)	1.54 (1.55)	1.39 (1.34)
	Total harvest loss	**6.33** (2.05)	**6.30** (2.22)	**7.12** (2.72)	**6.65** (2.42)
B	Threshing, winnowing, cleaning and drying	2.98 (1.92)	2.7 (2.03)	3.06 (2.09)	2.89 (2.05)
	Bulk handling for storage (Bagging and sacking)	0.86 (1.09)	0.65 (0.93)	0.94 (1.22)	0.80 (1.09)
	In-store (biotic & abiotic storage)	3.68 (2.85)	3.80 (3.15)	4.12 (3.75)	3.92 (3.38)
	Out-store (Destoring, bagging, sacking, transporting & marketing)	2.05 (1.39)	1.75 (1.44)	2.04 (1.52)	1.91 (1.47)
	Total post-harvest loss	**9.57** (1.81)	**8.90** (1.89)	**10.16** (2.15)	**9.52** (2.00)
C	Total wastage (A+B)	**15.90** (1.93)	**15.20** (2.05)	**17.28** (2.43)	**16.17** (2.21)
D	Seed	**2.49** (3.31)	**1.85** (2.29)	**1.24** (2.51)	**1.58** (2.68)
E	Feed (Cattle, poultry, goat and pet animals)	**3.49** (4.28)	**2.26** (2.81)	**1.39** (2.40)	**1.89** (2.78)
F	**Total seed, feed, wastage (SFW), (C+D+E)**	**21.88**	**19.31**	**19.91**	**19.64**

Figures in the parentheses indicate standard deviations
Source: Own calculation

Farm category wise total wastage (harvest and post-harvest losses)

The farm category wise estimation of seed, feed and wastage is presented in Table 15. Total harvest losses for large, medium, small and marginal farm were 6.99%,

6.72%, 6.68% and 6.39% respectively. The highest loss was incurred in large farms and the lowest loss was in marginal farms. Total post-harvest losses for large, medium, small and marginal farms were 10.16%, 9.66%, 9.58% and 9.05% respectively, with the average total loss for all farms at 9.52%.

Tab. 15: *Farm category wise average seed, feed and wastage (SFW) of rice*

	Items	Average losses and SFW (%)				
		Large	Medium	Small	Marginal	Average
A	Harvest operations (Cutting, field drying & bundling)	5.51 (3.88)	5.27 (3.49)	5.31 (3.39)	5.1 (3.52)	5.26 (3.50)
	Transporting from field to farm yard/threshing yard	1.48 (1.43)	1.45 (1.44)	1.37 (1.28)	1.29 (1.26)	1.39 (1.34)
	Total harvest loss	**6.99** (2.05)	**6.72** (2.22)	**6.68** (2.72)	**6.39** (2.72)	**6.65** (2.42)
B	Threshing, winnowing, cleaning and drying	2.94 (1.92)	2.99 (2.12)	2.9 (2.03)	2.75 (2.02)	2.89 (2.05)
	Bulk handling for storage (Bagging and sacking)	0.76 (1.03)	0.81 (1.09)	0.78 (1.03)	0.84 (1.18)	0.80 (1.09)
	In-store (biotic and abiotic storage)	4.48 (4.35)	3.92 (3.10)	4.00 (3.28)	3.59 (3.36)	3.92 (3.38)
	Out-store (destor, bag, sack, transport and marketing)	1.98 (1.40)	1.94 (1.48)	1.90 (1.45)	1.87 (1.53)	1.91 (1.47)
	Total post-harvest loss	**10.16** (2.32)	**9.66** (2.07)	**9.58** (2.02)	**9.05** (1.93)	**9.52** (2.00)
C	Total wastage (A+B)	17.15 (2.18)	16.38 (2.12)	16.26 (2.31)	15.44 (2.38)	16.17 (2.21)
D	Seed	1.48 (5.72)	1.44 (2.49)	1.84 (1.35)	1.72 (1.23)	1.58 (2.68)
E	Feed (Cattle, poultry, goat and pet animals)	1.38 (2.32)	1.88 (2.62)	2.24 (2.92)	2.46 (3.56)	1.89 (2.78)
F	Total seed, feed, wastage (SFW), (C+D+E)	20.01	19.70	20.34	19.62	19.64

Rice type wise stock availability in a crop season
Source: Own calculation

Wastage for different farm categories varied from 15.44% to 17.15%. The seed kept by large, medium, small and marginal farms were 1.48%, 1.44%, 1.84% and 1.72% respectively with the average being 1.58%. The feed usage by large, medium, small and marginal farms were 1.38%, 1.88%, 2.24% and 2.46% respectively with the average being 1.89%.

Rice type wise average farm stock availability data are shown in Table 16. The table showed that average per farm 1036 kg *Aus* rice (about 57%), 1174 kg *Aman* rice (about 29%) and 1291 kg *Boro* rice (about 21%) was available for meeting all family requirements.

Tab. 16:   Rice type wise average stock availability in a crop season

	Items	Per farm average amount in kg			
		Aus	Aman	Boro	average
A	Gross production	1832.80	4083.88	6205.55	4505.07
B	Seed, feed, wastage	401.02 (21.88)	788.60 (19.31)	1235.53 (19.91)	884.80 (19.64)
C	Net production (A-B)	1431.78 (78.12)	3295.28 (81.62)	4970.02 (80.09)	3620.27 (80.36)
D	(i) Consumption requirement	974.97 (53.20)	1065.95 (26.10)	1133.14 (18.26)	1079.83 (23.97)
	(ii) Donations	60.93 (3.32)	108.3 (2.65)	158.05 (2.55)	121.65 (2.70)
	Household use, (i)+(ii)	1035.9 (56.52)	1174.25 (28.75)	1291.19 (20.81)	1201.48 (26.68)
E	Marketable surplus, (C-D)	395.88 (20.60)	2121.03 (51.94)	3678.83 (59.28)	2418.79 (53.69)
F	Sale	596.30 (32.53)	2051.71 (50.24)	3765.04 (60.67)	2423.83 (53.80)
G	Purchase	346.22 (18.89)	141.38 (3.46)	195.51 (3.15)	187.17 (4.15)
H	End stock availability (E-F+G)	145.80 (7.96)	210.70 (5.16)	109.30 (1.76)	182.13 (4.04)

Figures in the parentheses indicate percentage per farm gross production
Source: Own calculation

All family requirements include consumption requirement and donations to others. Total average amount of all rice available per farm was about 1201 kg (about 27%). The household use of rice was obtained from net production which was derived from deducting seed, feed and wastage from gross production. Seed, feed and wastage (SFW) were about 22% for *Aus* rice, 19% for *Aman* rice and 20% for *Boro* rice with total average SFW of all rice being about 20%. After meeting SFW and household use, a portion of rice was obtained as marketable surplus. The per farm marketable surplus of *Aus* rice, *Aman* rice and *Boro* rice were respectively about 21%, 52% and 59% of gross production. Per farm total average marketable surplus for all rice was about 54%. Per farm marketable surplus from net production for *Aus* rice was about 27%, *Aman* rice was 64% and *Boro* rice 74%. The total average per farm marketable surplus was about 67%. From the marketable surplus, farmers sold considerable amount of their produced rice at different times in different crop seasons. Sometimes they sold more than surplus, which did shortfall their house requirements. Later on, they purchased some amount to fulfil the household needs or additional stock. At the end of each crop season, the stock availability per farm was about 8% for *Aus*, 5% for *Aman* and about 2% for *Boro* with total per farm average of 4%.

Overall average SFW and net production

The overall average total seed, feed and wastage were 1.58%, 1.89 and 16.17%. The net production again constitutes the summation of marketable surplus (53.68%) and the household uses (26.68%).

Potential production losses and SFW

It is necessary to provide knowledge about total rice losses both in pre and post production of rice as rice is the staple food for our population.

Tab. 17: Rice type wise total losses including potential production loss

Rice type	SFW (%)	Potential loss (%)	Total (%)
Aus	21.88	4.0	25.88
Aman	19.31	11.0	30.31
Boro	19.91	14.0	33.91
All	19.64	12.0	31.64

Source: Own calculation

Estimated total losses of rice were the highest for *Boro* (33.91%) followed by *Aman* (30.31%) and *Aus* (25.88%) respectively with the average total loss for all rice at 31.64%. Potential production loss due to technical inefficiency effects was the highest for *Boro* rice followed by *Aman* and *Aus* rice respectively (Table 17).

Conclusions, Policy Implications, Recommendations

Conclusions

Rice plays a dominant role in providing food for the people of Bangladesh. Rice is produced almost all over the country. *Boro* rice exhibited the highest productivity, production cost, gross return, net return per hectare and benefit cost ratio (BCR) followed by *Aman* and *Aus* rice. Productions of all rice crops were profitable. *Aus* and *Aman* were technically efficient but *Boro* was technically inefficient. Harvest and post-harvest losses were the highest for *Boro* rice and lowest for *Aman* rice. Per farm net production was the highest for *Boro* rice but the lowest for *Aus* rice. Marketable surplus was the highest for *Boro* rice (59.28%). Seed-feed-wastage (SFW) was the lowest in *Aman* rice (19.31%) and the highest in *Aus* rice (21.88%). Total losses of rice including potential loss were the highest for *Boro* (33.91%) followed by *Aman* (30.31%) and *Aus* (25.88%) respectively with the average total loss for all rice at 31.64%. Production of rice was affected by technical efficiency factors. Its availability was affected by the harvest and post-harvest losses.

The largest portion of the total produce was kept for home consumption. Other large share of production distributed was donations followed by non-human food usages. Farmers used to sell a portion of rice especially of *Boro* rice for meeting their cash needs. On an average, per farm end stock availability remained only 182 kg (4.04% of the total gross production). For reducing SFW and increasing farm stock availability, farmers suggested several ways. Some of the ways were timely cutting, applying efficient harvest machine, threshing of rice in the protected ground in the field, in a common field as far as possible by power thresher, constructing village level common storage structures along with training and/or providing loan facilities for constructing individual safe storage structures.

Policy implications - Long term policy options

Aus and *Aman* were technically efficient, further increased production cannot be achieved by increasing efficiency of these crops. For these rice types advanced technology is inevitable to increase production. Although *Boro* rice production can be increased with the existing variety, advanced technology is essential for this crop in the long run. Therefore, research organisations are to be well equipped in order to develop higher yield promising and drought tolerant varieties. Research can also be carried out to develop less grain shattering varieties to reduce harvest loss.

Rural infrastructures with connecting routes should be further developed for easy access and carrying rice from field to farmyard and farmyard to market. The motivation for the rural infrastructures development is to reduce transportation loss and also to increase access of farmers to markets. Government can take initiative to develop rural infrastructures.

Policy implications - Short term policy options

Some agricultural lands remained fallow due to lack of irrigation facilities. These lands could be brought under cultivation to increase total volume of output and stock availability. Irrigation facilities should be developed to bring more land under rice production. Uninterrupted electricity supply should be ensured during the season of *Boro* rice production.

Recommendations

- Government rice procurement programmes should be revised to ensure farmers' access to sell their rice directly at their farm levels or local markets.
- Rice should be harvested at optimum maturity with appropriate harvest technology to reduce harvest loss. Measures should be taken to make available the harvest technology to farmers.
- Farmers should be encouraged to perform post- harvest operations in common yard in the field wherever possible.
- Farmers should be encouraged for threshing, winnowing and drying their paddy with mechanised thresher on confined and protected common paved floor.
- Appropriate storage structures should be developed to reduce storage losses. Public extension organisations should come in close contact with the farmers for effective use of these storage structures.
- Farmers' awareness programmes on capacity building about harvest, post- harvest including proper storage should be implemented.

Note

This study has been conducted under the Programme Research (PR) grant received from National Food Policy Capacity Strengthen Programme, FAO (GCP/BGD/034/MUL). The researchers are thankful to FAO for funding this project.

List of References

Ahmed, R. (2001): Restrospects and prospects of the rice economy of Bangladesh. The University Press Limited, Dhaka.

Alam, M.S., Ali, M.A., Mia, M.I.A. and Abedin, M. Z. (2007): Studies on grain storage facilities as food security measure in flood prone areas of Bangladesh. The Journal of Progressive Agriculture, 18 (2), 223-233.

Banglapedia (2008): CD Edition, February 2008. National Encyclopedia of Bangladesh. Published by Asiatic Society of Bangladesh.

Banik, A. (1994): Technical efficiency of irrigated farms in a village of Bangladesh, Indian J. Agricultural Economics, 49 (1), 70-78.

BBS (2009): Statistical Yearbook of Bangladesh 2008. Ministry of Planning. Government of the Peoples' Republic of Bangladesh. Dhaka.

Calverley, D.J.B (1994): Programme for the prevention of food losses: A study of eleven projects in Asia concerned with rice. Final report, FAO.

Deb, U. K. and Hossain, M. (1995): Farmers' Education, Modern Technology and Technical Efficiency of Rice Growers. Bangladesh J. Agric. Econ. 18 (2), 1-13.

FAO (2004): Post-harvest losses: Discovering the full story. FAO Corporate Document Repository. Available at www.fao.org/DOCREP/004/AC301E/AC301e04.htm.

FAO (2010): Food Outlook: Global Market Analysis, Global Information and Early Warning System, An internet version.

Mandoza, T.L and Gummert, M. (2007): Fighting Asia's Post-harvest Problems, Rice Today-International Magazine, IRRI.

NAPHIRE (1997): Technical Guide on Grain Postharvest Operation. National Postharvest Institute for Research and Extension (NAPHIRE). Munoz, Nueva Ecija, Philippines.

Quasem, M.A. and Siddique, M. A. (2009): Reduction of Post-harvest losses in Rice, A paper presented at CIRDAP Workshop, on Nov. 23, CIRDAP, Dhaka.

Rahman, K. M. M., Schmitz, P. M. and Wronka, T. C. (1999): Impact of Farm-Specific Factors on the Technical Inefficiency of Producing Rice in Bangladesh. Bangladesh J. Agric. Econs. 22 (2), 19-41.

Rahman, K.M.M. (2002): Measuring Efficiency of Producing Rice in Bangladesh: A Scochastic Frontier Analysis. Wissenschaftsverlag Vauk, Kiel.

Rahman, S.M.M. (2009): Rice Postharvest Situation and Strategy for the Reduction of Losses in Bangladesh, A paper presented at CIRDAP Workshop, on Nov. 23, CIRDAP, Dhaka.

Samajpati, J.N. and Sheikh, S.A. (1980): Paddy and rice storage in Bangladesh with emphasis on insect infestation. Agr. Mech. In Asia, 11 (1), 69-72.

Climate Change Impacts on Agriculture and the Relevance of Adaptation: The Case of Pakistan

Mirza Nomman AHMED

Abstract

This study uses the ricardian climate change valuation technique to estimate the economic impacts of climate change on agriculture in Pakistan. Existing studies of climate change impacts for the country do not consider the full set of farmer adaptations and thus tend to overestimate climate change induced losses. This study contributes to the literature by addressing this shortcoming and additionally providing an analysis of spatial effects across Agro-Ecological Zones, provinces and farm-types using a multi-seasonal approach. Departing from an intelligent farmer's scenario, where adaptation has ever since taken place, a hedonic analysis of farmland pricing, with climate as one determinant, is carried out. To assess the future impacts, a response function is estimated and simulations are run in a steady-state setting using Global Circulation Model scenarios. Based on the possession of agricultural land 22240 owner operated farms are selected and aggregated into 110 districts. The results confirm a non-linear relationship between climate and agricultural practice. Rainfed farms are found to be more vulnerable, whereas additional irrigation seems to be an option for farmers. All Agro-Ecological-Zones and Provinces suffer from further warming in the summer season, whereas benefits are revealed for the winter season. Those zones most severely hit, also have the lowest land values per hectare, precisely under 500$. When warming is simulated further, interesting results are revealed. Winter season benefits by 2090 over-compensate summer season losses, however not in the dryer parts of the country. In general, farmers in Pakistan are found to have well adapted to the current climatic conditions. Given climate change and if nothing else changes, they would not have to suffer great losses.

Introduction

Pakistan is a developing country facing several obstacles to development, including climate change and climate variability. In a nutshell, thus, vulnerability to climate change can be framed as follows: the climate is arid and hot, regularly the country is exposed to risks from extreme weather events (droughts and floods), socio-economic conditions are deteriorating, major income sources of the population (>170 million) depend on climate sensitive sectors, awareness for environmental protection is missing, warning systems are not in place or are outdated, political instability is of concern and capacity to react to abrupt and long-term environmental changes is low. Specifically the agricultural sector is at risk as it is the mainstay of the population, with a total contribution of 25% to overall GDP. The industry of the country is heavily agro-based and almost two thirds of agriculture are irrigated. More than 30% of the population live below the poverty line. GNI per capita is estimated at 3030 $ (PPP) (WDI, 2013). The frequency and the intensity of climate related extreme events in Pakistan have both increased in the recent past. From 1998-2002 the province of Balochistan was hit by severe drought conditions, affecting 84% of the population directly, killing 76% of the province's livestock and causing mass migration due to widespread hunger and disease. In 2010, 84 of 121 districts were severely effected by epic floods. According to the climate change vulnerability index Pakistan is ranked 12th globally and total economic losses of approximately 4.5 billion dollars are anticipated in future, owing to significantly higher temperatures and decreased surface water availability and changing precipitation patterns (LP, 2008). However, all these future estimates of losses and damages for Pakistan do not consider implicit adaptation by the farmers (e.g. through crop switching) in their calculations. These adaptations have taken place since the existence of agriculture and will continue to play a role in the wake of a shift in agro-climatic conditions (KUSTERS and WANGDI, 2013). Hence, failure to consider the full set of farmer adaptations will lead to overestimations of the damages from climate change. Despite an internationally extensive interest in the measurement of the economic impacts of climate change, specifically on agriculture, the empirical research on Asia remains scarce.

Although models have been constructed to assess the economic impact of global warming on agriculture, a clear consensus on the methodology and impacts has not yet evolved. According to MENDELSOHN (2000) the literature so far (not much has changed) suggests that over the next hundred years global food supplies in aggregate will not be harmed. In a detailed study for the FAO, MENDELSOHN (2000) also states that "...warming is not expected to affect aggregate production in most developing countries". However, at the same time results suggest that productivity declines will be inevitable, especially in regions where temperatures are expected to increase and precipitation is predicted to decline. Using an extensive literature survey WASHINGTON (2006) found that Asia alongside Africa, the Middle East and South America has been a neglected region as far as climate research is concerned. They concluded, that this shortcoming is detrimental to climate risk assessment, planning of adaptation and decision making in developing countries. This study's mandate is to partially address these shortcomings by using the Ricardian valuation approach developed by MENDELSOHN et al. (1994) as a basis. A unique model is constructed which is adapted to the country's socio-economic and environmental circumstances. For this farm level data is aggregated. The analysis focuses on estimating the impacts of climate change on agriculture, also comprising their seasonal and spatial spread (Agro-Ecological Zones and provinces). Last but not least, in order to analyze the future impacts of climate change on Pakistan's agriculture, and to assess the importance of past adaptations, simulations using GCM (Global Circulation Model) scenarios are run.

Climate Change Vulnerability and the importance of Adaptation

Table 1 based on the CRED's (Collaborating Centre for Research on the Epidemiology of Disasters) data on emergency events presents a timeline of documented major extreme events in Pakistan, dating back as far as 1935 when Pakistan was known as Western India. Climate related natural hazards among others can comprise tornados, hurricanes, floods, cyclones, drought, landslides and heat or cold waves (CBSE, 2006). Pakistan however, has been specifically vulnerable to floods, storms and droughts (AHMED and SCHMITZ, 2011).

Tab. 1: Climate related natural hazards in Pakistan (1935-2011)

Type of Disaster	Date	Death Toll	No Total Affected	Damage (000 US$)
Earthquake	31.05.1935	60000	-	-
	27.11.1945	4000	-	-
	28.12.1974	4700	-	-
	08.10.2005	73338	5128309	5200000
Flood	1950	2900	-	-
	Aug 1973	-	4800000	661500
	02.08.1976	-	5566000	505000
	Jun 1977	848	-	-
	Jul 1978	-	2246000	-
	15.07.1992	-	6184418	-
	08.09.1992	1334	6655450	1000000
	02.03.1998	1000	-	-
	22.07.2001	-	-	246000
	09.02.2005	-	7000450	-
	10.08.2007	-	-	327118
	02.08.2008	-	-	103000
	28.07.2010	1985	20359496	9500000
	Aug 2011	-	5800000	-
Storm	15.12.1965	-	10000	-
	26.06.2007	-	-	1620000
Drought	Nov 99	-	2200000	247000

Source: EM-DAT VERSION 12.07 (2012)

According to the Pakistan Meteorological Department (CHAUDHRY et al., 2009) during the period 1901-2000, the increase in mean annual temperature in the Northern half of Pakistan was found to be higher with a value of 0.8°C as compared to the country as a whole with 0.6°C. Furthermore, data from the Climate Research Unit (CRU) in the United Kingdom relative to the national scale indicate a higher increase in mean annual temperature for Northern Pakistan.

In addition, based on data from 1951 to 2000 the TFCC (Task Force on Climate Change) report highlights the general warming trend in mean and maximum temperatures for the summer season (April and May), this throughout the country. For the same time period, the Monsoon Season that spans from July to September has generally shown a decreasing trend in temperatures (TFCC, 2010). According to McSWEENEY et al. (2010) the warming trend is primarily observed for the months from October to December, with precisely 0.19°C per decade. Additionally, the frequency of hot days and nights (days or nights where temperature exceeds a certain threshold by 10%) in Pakistan has increased, whereas the frequency of cold days and nights has decreased respectively. As far as precipitation is concerned, an increasing trend for precipitation over the last century is revealed. Although the frequency of fluctuations is large the extremes are most evident. As the majority of the annual precipitation is received in the monsoon period, this general rising trend can be explained by the increased variability of the monsoon.

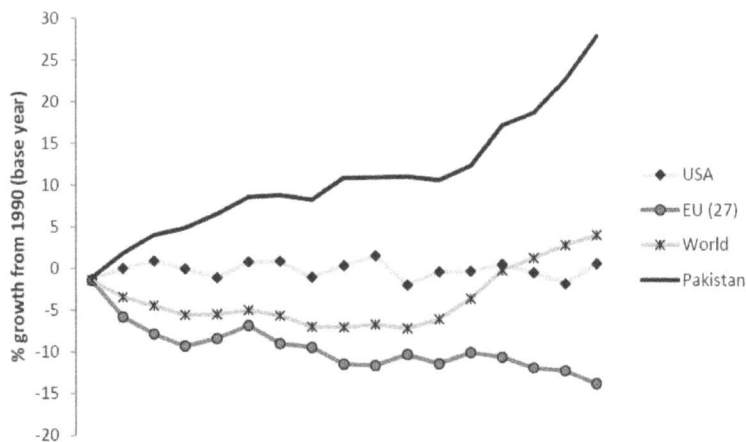

Fig. 1: Per Capita CO_2 Emissions in selected regions of the world, 1990-2007
Source: AHMED, 2013, p. 20 (after CAIT 8.0 (CAIT, 2011))

Despite the fact that the country's share in global GHG emissions is marginal, there is enough reason for concern in future when having a look at the annual percentage growth rate of per capita carbon dioxide emissions from 1990 to 2008 (Figure 1). In 2008 out of 309 million tonnes (mt) of carbon dioxide (CO_2) equivalent total Greenhouse Gas (GHG) Emissions, 39%, were contributed by the agricultural sector (CAIT, 2011; TFCC, 2010).

Besides the primary gas associated with global warming "carbon dioxide" (CO_2) three other factors are primarily important for crop or plant growth, these include sunlight, nutrients and water (WOLFE and ERICKSON, 1993; HOUGHTON et al., 2001). Figure 2 reveals that every species, be it crop or livestock, has an optimal range in which its growth is maximized. In this regard, the climate ranges sup-optimal and optimal can be distinguished. Certain so called limiting factors such as soil and water can impact the climate-crop relationship and either in case of the existence of a limitation lead to lower growth rates and in the absence of such to a higher growth rate, respectively. The existence of limiting factors such as degraded soils and water scarcity lead to overall lower levels of growth and production, this throughout all climate ranges. Most importantly though Figure 4 reveals that higher temperatures above the optimal threshold in combination with the existence of the aforementioned limiting factors are most harmful to crop and animal production. Farmers however are found to adjust their practices in order to restrict the impact of these above mentioned limiting factors.

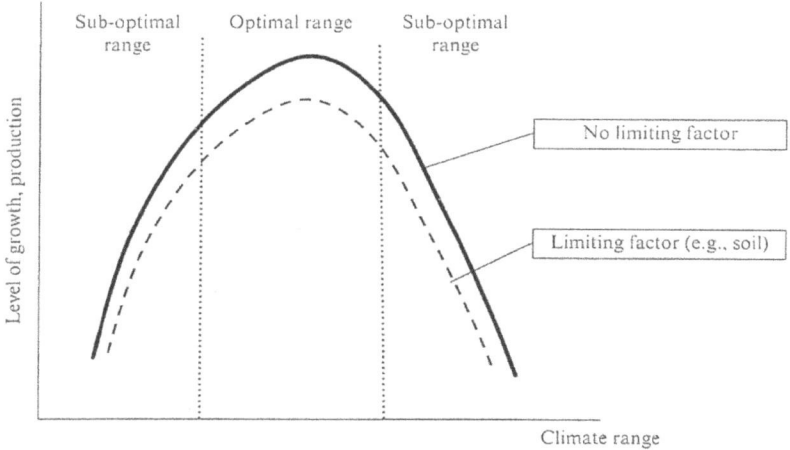

Fig. 2:   *Climate and soil interaction – impacts on crops and livestock*
Source: MENDELSOHN and DINAR, 2009

Although many different definitions exist, adaptations in the climate change context can be understood as measures undertaken to reduce the adverse impacts from global warming on natural and human systems. Individuals, firms, communities and other entities take these actions mainly for two reasons, either to reduce the damages or increase the benefits from climate change (IPCC, 2007; BARRY et al., 2000).

Local traditional knowledge has ever since the existence of human civilizations played a key role in guaranteeing a successful adaptation to ever changing conditions that humans have been faced with. A recent report under the umbrella of the CGIAR documents several local adaptation measures that famers undertake in the Indo-Gangetic Plains (RIVERA-FERRE et al., 2012). Table 2 documents the major uses of the indigenous knowledge that have been identified by the expert groups. As table 2 shows, a wide range of agriculture and food related activities are covered. From the climate context, a very interesting finding is the existence of local weather forecasting systems.

Tab. 2: *Major uses of local traditional knowledge in the Indo-Ganges Plains*

Management	Soil management
	Water management
	Terrace management
	Rain fed cropping
	Crop diversification
	Changes in cropping pattern
	Resource efficient cropping systems
Seeds	Seed saving
	Knowledge about what crop varieties to plant
	Local communities choice of varieties
	Local seed selection process, storage, exchange system
	Changes in crop varieties
Weather	Climate prediction (local weather forecasting system)
Food	Food preservation
	Food preparation/consumption

Source: RIVERA-FERRE et al., 2012

Importantly, farmers or communities also practice crop diversification, change cropping patterns and varieties, and manage water systems. Various traditional systems exist in Pakistan that are still used today.

Methodological Approach

To account for a full range of farmer adaptations when assessing the impacts of climate change on agriculture, the so called "Ricardian Approach" has been developed by MENDELSOHN et al. (1994). The Ricardian Model was named after 19th century economist David Ricardo, because of his observation that land rents would reflect the net revenue of farmland (RICARDO, 1817). It is not to be confused with the Ricardian Trade model. SCHLENKER et al. (2005) have identified the Ricardian model as a hedonic approach of farmland pricing. Climate change is a shift in the long-run pattern of weather over time, whereas weather occurs at a point in time with great fluctuation. Reference is made towards two different phenomena (weather and climate), thus the respective economic implications amongst them will also differ. To this background the system of measurement, dependent variable or economic metric will also have to be chosen accordingly. Especially for analyzing long-term phenomena like climate change, a measure is needed that reflects the permanent income from farming. "Land value" in this regard comes in handy as it reflects long-term profitability of using the land for farming (FISHER et al., 2007). Generally, the hedonic relation between land values and climate as an environmental variable arises due to the heterogeneity in the different qualities of land (ROSEN, 1974). These land quality attributes include climate amongst others and being hedonic in nature the model covers a wide range of substitution options (GOULDER and PIZER, 2006). Farmers in different regions are exposed to different climatic conditions, thus the Ricardian method by making use of cross-sectional data and spatial comparison assumes that they adapt to their local circumstances. As far as adaptation to climate change in this context is concerned, the core assumption of the Ricardian Approach is that if growing rice is more profitable than growing wheat, and if the climate becomes more suitable for rice than wheat, then those farmers will adapt to the changed climate by drawing on the experiences of rice farmers elsewhere and switching from wheat to rice (POLSKY, 2004).

The model starts from the observation that farmer decisions are dependent on uncontrollable exogenous factors including climate and that the "naive farmer" or "dumb farmer" scenario, in which the farmers regardless of occurring changes doggedly continue with their known practices, does not hold. HELMS et al. (1996) argue with the slow nature of climate change, where time is given to the farmers to react and adapt. Thus given these exogenous constraints a farmer has to choose a certain mix of inputs (e.g. crops species, pesticides, irrigation) to maximize his profits. In a cross-sectional framework a comparative steady state analysis of long-term climate impacts on land values is carried out.

The Ricardian model starts with the assumption of a set of well-behaved production functions.

$$Q_i = Q_i(K_i, E) \qquad i = 1, \ldots\ldots, n \qquad (1)$$

Where, the vector Ki denotes the purchased inputs for the production of a specific good i, K_{ij} is the purchased input j for the production of good i and the vector E stands for external environmental inputs including climate (location specific soil characteristics, precipitation and temperature). As the inputs, environmental factors and the output Q_i have prices w_j, cost minimization leads to a cost function of the following form:

$$C_i = C_i\ (Q_i, w, E\ ) \qquad (2)$$

With C_i denoting the production costs of good i and the vector w representing the factor prices. Now, the assumption of utility maximizing consumers with well behaved utility functions and linear budget constraints, who take prices as given, leads to a system of inverse demand functions for the outputs i=1, n of the following form (SANGHI, 1998):

$$P_i = D^{-1}\ (Q_1, \ldots Q_i, \ldots Q_n, Y\ ) \qquad (3)$$

Where, Y is the aggregate income, P_i denotes the price of good i and Q the quantity of the good. Profit maximization yields:

$$\max_{Q_i} P_i\ Q_i - C_{i-}(Q_i, w, E) - p_L L_i \qquad (4)$$

With p_L as the annual cost of land at a site and $C_{i-}(Q_I, w, E)$ as the cost function of all purchased inputs except land; with $C_{i-}(Q_I, w, E) = C_i + p_L L_i$.

Assuming perfect competition in the land markets drives profits to zero yielding the following equation:

$$P_i Q_i^* - C_{i-}{}^*(Q_i^*, w, E) - p_L L_i^* = 0 \qquad (5)$$

If under the given environmental conditions use i is the best use of land, the net revenues from the production of good i will equal the observed market rent.

In other words, land rent per hectare is equal to net revenue per hectare; this is captured in the following equation, which is solved for p_L.

$$p_L = [P_i Q_i^* - C_{i-}^*(Q_i^*, w, E)]/L_i \tag{6}$$

Land value reflects the present value of the future stream of income from a parcel of land. A long term accumulation of net revenue determines farmland value (MENDELSOHN and NORDHAUS, 1996).

$$V_L = \int_0^\infty p_L\, e^{-\varphi t}\, dt = \int_0^\infty [P_i Q_i - C_{i-}(Q_i, w, E)]\, e^{-\varphi t}/L_i dt \tag{7}$$

or

$$V = \beta_0 + \beta_1 F + \beta_2 F^2 + \beta_3 Z + \beta_4 G + u$$

Where t is time, φ is the discount rate and L_i denotes the amount of land in farm i. Land Values (V) are affected by the set of exogenous variables F, Z, and G. F is the vector of climate variables, Z is the sets of soil classification, G is the set of socio-economic indicators. Equation (7) has been termed the "essence of the Ricardian model" by MENDELSOHN and NORDHAUS (1996) (Figure 3).

Fig. 3: *The essence of the Ricardian Model*

Source: Own presentation after MENDELSOHN and NORDHAUS (1996)

As the main goal is to value climate change, the change in climate from an initial level E_A to a new level E_B also reveals the change in annual welfare, which is given by:

$$\Delta W = W(E_B) - W(E_A) = \int_0^{Q_B} \Sigma D^{-1}(Q_i) dQ_i - \Sigma C_{i-}(Q_i, w, E_A) -$$

$$\int_0^{Q_A} \Sigma D^{-1}(Q_i) dQ_i - \Sigma C_{i-}(Q_i, w, E_B) \tag{8}$$

Assuming constant market prices over both climate levels leads to a reduced equation:

$$\Delta W = W(E_B) - W(E_A) = PQ_B - \Sigma C_{i-}(Q_i, w, E_A) -$$

$$[PQ_A - \Sigma C_{i-}(Q_i, w, E_B)] \qquad (9)$$

Figure 4 illustrates the welfare impact of a change in the environmental variable from an initial state A to a new level B, where the supply curve (marginal cost curve) shifts outwards. A substitution of the land per hectare equation into equation (9) gives:

$$\Delta W = W(E_B) - W(E_A) = \Sigma_i(p_{LB} L_B - p_{LA} L_A) \qquad (10)$$

Where the land rent per hectare in the initial climate state A (E_A) is subtracted from the land rent per hectare in the new climate state B (E_B). The following equation (EQUATION 11) gives the present value of this welfare change.

$$\int_0^\infty \Delta W e^{-\varphi t}\, dt = \Sigma_i(V_{LB} L_B - V_{LA} L_A) \qquad (11)$$

The present value of the stream of welfare change is captured in the difference between land values in the new environmental state and land values in the initial environmental state. Dependent on the choice of the dependent variable, whether it is annual net revenue or farmland value (capitalized net revenues) the Ricardian Model can take the form of either equation (6) or equation (11). Thus, by using cross-sectional observation a change in environmental conditions (normal climate) is reflected in land values and production. As land values or net revenues vary with location and climatic conditions, farmer adaptation is implicitly captured. A beneficial climate change is embodied in higher land values, whereas if farmland values decrease, the climate change is harmful. The aggregate effect is found by summing up all effects over all observations and applying weights (e.g. amount of land in a farm, region or district). The model can thus be estimated for districts, regions and Agro-Ecological Zones. After estimating the so called Ricardian climate response function with current climatic conditions, one can replace the current climate situation with any other climate change scenario and simulate future impacts. However, as the ability of farmers to adapt besides environmental changes is also bound to many other factors that may well change over time (e.g. human capital, government support structure, technology) (MENDELSOHN and DINAR, 2009) have clearly called for a cautious approach when interpreting the results of the simulations and future predictions based on the Ricardian estimates of climate change impacts on agriculture. To sum up, the Ricardian valuation model's key hypothesis is that through a change in climate the production function (crops or livestock) will be shifted. As farmers observe these changes in climate, they manage to adapt to their local conditions by adjusting their inputs and outputs accordingly. This implies that in the current situation, the economy has well adapted and land prices have reached a long-run equilibrium in the different districts (with climate as a determining factor). Various countries have been studied using the Ricardian Approach (Figure 4).

Note: only peer reviewed publications have been considered following MENDELSOHN and DINAR (2009)

Fig. 4: Global coverage of the Ricardian Approach 1994-2011

Source: AHMED, 2013, p. 117

Discussion of Results

The analysis was conducted using a step-wise approach, where modeling started with only climate variables and was augmented with other relevant control variables. In total three separate models were estimated for the available dataset. First, a model was estimated for the entire sample, covering 109 districts. To test whether the impacts of climate differ for rainfed and irrigated farming systems, two additional models were estimated, covering 77 and 109 districts, respectively. Several different econometric specifications were examined before settling on the specification in table 4.

The second and third columns of the table show separate regressions for rainfed and irrigated farms. The different specifications included a two-seasonal, a three-seasonal and a four-seasonal model. The first major specification included the agricultural growing seasons of the country, namely Kharif (summer) and

Rabi (winter) only. In the second specification an additional season was added with spring, to test whether pre-sowing conditions are important for explaining the variation in land values. In the final specification a fourth season was added. The final specification thus was made up of a summer sowing (compares to spring), a summer growing (compares to summer), a winter sowing (compares to fall) and a winter growing season (compares to winter). By disaggregating the season into sowing and growing periods the model was prepared to test for specific features of climatic timing. The four-seasonal model with disaggregated sowing and growing periods seemed to capture the individual inter-seasonal effects best. In the next step the idea was to test the assumption whether season specific interactions between precipitation and temperature are important. The results proved not very promising and particularly led to a significant amount of instability in the model, mainly through substantially raising variance inflation (multicollinearity). Furthermore, the season-specific interaction terms were neither significant nor did they have the expected signs and thus were excluded. Instead, to test whether surface water availability and temperature interact, an interaction term between surface water (runoff and river density) was introduced. Additionally, the irrigated model added an interaction term between irrigated acreage and annual temperature. Adding these annual interaction terms instead of seasonal ones did not raise issues with multicollinearity and the results remained stable. The result indicated that these interactions are significant and important. Agronomic research suggests a hill-shaped relationship between temperature and crop growth (see first paragraph), however several other functional forms are possible (KAUFMANN, 1998). Positive linear and negative quadratic terms translate into a hill-shaped functional relationship, whereas vice versa the relationship is U-shaped. A hill-shaped relationship indicates that up to a certain level of increase, climate will have a beneficial effect on land values (or capitalized net revenues) but after having reached a maximum, land values will start to decline with an increase in the respective climate variable. In contrast to the hill-shaped relationship the U-shaped relationship between climate variables and land values indicates an initial decrease in land values with an increment in the climate variable, which however after reaching a minimum, again starts increasing.

As table 4 shows, most climate variables in all three models are significant. In general all three models confirm the high sensitivity of farming in Pakistan to climate. Detailed summary statistics and data sources are presented in table 3.

The results confirm the agronomic and agro-climatic wisdom of a non-linear relationship between climate and crop growth and ultimately farm incomes (land values), indicated by significant linear and squared coefficients. Higher temperatures in the early part of the respective farming season seem to be beneficial. Likewise, a hill-shaped trend for the later stages of the farming season (i.e. summer growing) may indicate a negative impact for farming. The proxy for solar radiation (Latitude) is not significant in the full model. General control variables such Human Development Index, population density, elevation etc. and more specific control variables such as soils and hydrology are mostly significant and have the expected signs, clearly indicating the importance of the inclusion of controls when assessing the impact of climate change in a hedonic setting. In the regression for rainfed farms, four out of eight temperature coefficients and four out of six precipitation coefficients are significant. In case of irrigated farms, six out of eight temperature coefficients and two out of six precipitation coefficients are significant.

Variable	Unit	Rainfed Farms				Irrigated Farms				Data Sources
		Mean	Std. Dev.	Min	Max	Mean	Std. Dev.	Min	Max	
Land Value	$/ha	1461	2024	30.73118	13391.06	2306.156	3352.872	30.07491	30506.6	FBS, 2010
Summer Sowing Temp	(°C)	29.014	4.6113	17.6	35.2	29.39764	4.501334	17.6	35.2	NCDC, 2011; GOMMES et al., 2004
Summer Growing Temp	(°C)	29.886	3.094	22.56667	35.36667	30.17031	2.980283	22.56667	35.36667	NCDC, 2011; GOMMES et al., 2004
Winter Sowing Temp	(°C)	17.999	4.0497	8	24.53333	18.31948	4.025999	8	24.6	NCDC, 2011; GOMMES et al., 2004
Winter Sowing Temp^2	(°C)	14.345	4.3467	5.233334	21.73333	14.7122	4.408709	2.833333	21.6	NCDC, 2011; GOMMES et al., 2004
Summer Sowing Precip	(mm)	89.82	104	5	506.1	77.29	92.97647	5	506.1	NCDC, 2011; GOMMES et al., 2004
Summer Growing Precip	(mm)	53.086	52.073	1.25	229.25	51.54764	47.21411	1.25	229.25	NCDC, 2011; GOMMES et al., 2004
Winter Precipitation	(mm)	25.83	27.467	1.333333	115	29.63807	32.75382	2	153.6667	NCDC, 2011; GOMMES et al., 2004
HDI	(HDI)	0.60371	0.09131	.31	.802	.6180275	.084598	.31	.802	JAMAL AND KHAN (2007).
Population Density	(No./km^2)	249.06	276.81	4	1606	340.6514	482.345	4	3566	PCO, 1998
Elevation	(m)	620.43	700.56	5	3314	554.8807	693.838	5	3314	ArcMap (ESRI, 2011); Google Maps
Latitude	(Decimal°)	30.513	3.1595	24.67	35.84	30.39505	2.945104	24.67	35.84	ArcMap (ESRI, 2011), Google Maps
Runoff	(mm/year)	53.076	123.31	0	627	42.56881	107.2055	0	627	BUDYKO (1974); LocClim 1.10 (FAO, 2005)
River Density	(km)	30.86	32.932	0	190.84	33.11679	32.35007	0	190.84	STRZPEK AND MCCLUSKEY'S (2006); ESRI, 2011
Flood	(% affected area)	0.12586	0.17828	0	.8287163	.1417012	.2107625	0	.973464	SUPARCO & FAO (2010)
Tractor Density	(No./ha)	1009.7	2177	0	9917.05	.0991071	.1264484	0	.6467662	RAFIQUE, 2009; GOS, 2008; GON, 2009; GOB, 2012
Road Density	(km/ha)	2.7055	4.6484	.004	38.51734	2.933349	4.1079	.004	38.51734	RAFIQUE, 2009; GOS, 2008; GON, 2009; GOB, 2012
Soil Calcisols	(%area)	0.35742	0.33659	0	1	.4353572	.3558461	0	1	FAO, 2003, 2006 (DSMW)
Soil Fluvisols	(%area)	0.0448	0.0953	0	.4134663	.0701113	.1188612	0	.5094377	FAO, 2003, 2006 (DSMW)
Soil Rock outcrop	(%area)	0.0342	0.13589	0	.738439	.0447469	.1676542	0	1	FAO, 2003, 2006 (DSMW)
Soil Regosols	(%area)	0.06253	0.19569	0	1	.0244784	.1157886	0	.738439	FAO, 2003, 2006 (DSMW)

Tab. 3: Summary Statistics and Data Sources (Source: Own calculation)

Climate Change Impacts on Agriculture and the Relevance of Adaption

Tab. 4: *Ricardian Climate Regressions- Estimation Results*

Variables		Total Farms	Rainfed Farms	Irrigated Farms
Land Value/acre (in Rs.)		*Climate Variables*		
Summer Sowing Temperature	(°C)	0.0192	-0.0630	0.0404
		(0.0344)	(0.0637)	(0.0386)
Summer Sowing Temperature^2	(°C)	0.0195***	0.0189***	-0.00728***
		(0.00419)	(0.00677)	(0.00276)
Summer Growing Temperature	(°C)	-0.110***	-0.0769**	-0.168***
		(0.0217)	(0.0346)	(0.0506)
Summer Growing Temperature^2	(°C)	-0.00474*	-0.00425	0.0108**
		(0.00258)	(0.00408)	(0.00429)
Winter Sowing Temperature	(°C)	0.241***	0.300***	0.158**
		(0.0662)	(0.112)	(0.0687)
Winter Sowing Temperature^2	(°C)	-0.00506	-0.00330	-0.0181**
		(0.00744)	(0.0147)	(0.00697)
Winter Growing Temeprature	(°C)	-0.216***	-0.239*	-0.103
		(0.0684)	(0.120)	(0.0655)
Winter Growing Temperature^2	(°C)	-0.0172***	-0.0171	0.0113*
		(0.00597)	(0.0111)	(0.00610)
Spring Precipitation	(mm)	-0.000393	0.00710*	0.000805
		(0.00232)	(0.00388)	(0.00243)
Spring Precipitation^2	(mm)	-2.62e-05***	-5.43e-05***	-9.42e-06
		(9.48e-06)	(1.58e-05)	(5.77e-06)
Monsoon Precipitation	(mm)	-0.0146***	-0.0177***	-0.0103***
		(0.00172)	(0.00309)	(0.00173)
Monsoon Precipitation^2	(mm)	3.55e-05***	1.91e-05	1.31e-05
		(1.24e-05)	(2.08e-05)	(1.25e-05)
Winter Growing Precipitation	(mm)	0.000447***	-0.00263	0.0177***
		(0.000166)	(0.00895)	(0.00612)
Winter Growing Precipitation^2	(mm)	1.30e-07**	0.000360***	5.88e-05
		(6.47e-08)	(0.000110)	(5.21e-05)
Latitude	Decimal Degrees	0.0481	-0.0819	-0.0468
		(0.0338)	(0.0628)	(0.101)
		General Control Variables		
Human Development Index	HDI	1.696***	1.551**	2.943***
		(0.364)	(0.583)	(0.410)
Population Density	Population/km^2	0.0323	0.00100***	0.270***
		(0.0342)	(0.000262)	(0.0354)
Elevation	m	-0.189***	-0.150**	0.197***
		(0.0388)	(0.0649)	(0.0434)
Tractor Density	tractors/ha	0.116***	0.108***	0.0488**
		(0.0182)	(0.0306)	(0.0194)
Road Density	road km/area	0.0495***	0.0359***	0.0311***
		(0.00828)	(0.0133)	(0.00901)

continued.........		Hydrology and Interactions		
Runoff	mm	0.199***	0.213***	0.0195*
		(0.0349)	(0.0593)	(0.0113)
River Density Index	River lenght km/area	0.0296**	0.0498*	0.0135
		(0.0129)	(0.0249)	(0.0297)
Flood Risk	% of total area affected	-0.496***	-0.697*	-0.131
		(0.141)	(0.375)	(0.148)
Surface*Temperature	mm*(°C)	-0.000303***	-0.000273**	-6.96e-05
		(7.36e-05)	(0.000123)	(8.99e-05)
Irrigation*Temperature	ha*(°C)	-	-	3.73e-06*
				(2.17e-06)
		Soil Type		
Soil Calcisols	% in total area	0.337***	-0.346	0.237**
		(0.115)	(0.256)	(0.117)
Soil Fluvisols	% in total area	0.588*	0.946	0.652**
		(0.325)	(0.736)	(0.324)
Soil Rock Outcrop	% in total area	0.316*	0.541*	0.494**
		(0.183)	(0.294)	(0.196)
Soil Regosols	% in total area	-0.163	1.017**	-0.865***
		(0.299)	(0.483)	(0.321)
Constant		12.12***	11.65***	9.791***
		(0.311)	(0.483)	(0.335)
		Robust regression measures of fit		
Observations		109	77	109
R-square		0.69882549	0.70607471	0.7039479
AICR		224.48602	153.31708	219.75103
BICR		318.68092	235.55413	315.25755
deviance		5.2503218	5.9265026	5.4961853

Standard errors in parentheses *** p<0.01, ** p<0.05, * p<0.1
Dependent Variables is logarithm of Land Value/acre in Rupees (80 Rupees=1$)
Note: the respective season's sowing period represents the opposite season's harvesting period!

Source: Own calculation

Both disaggregated models have the expected sign for the solar radiation proxy (latitude). A detailed interpretation of the control variables will be presented in the upcoming sections.

The interpretation of the climate coefficients in raw format is rather difficult as linear and quadratic terms are included. These terms can however indicate totally different functional relationships between climate and agricultural land values. To this background, in order to get an unmitigated understanding of the results regarding climate, in following the marginal impacts or so-called Ricardian climate sensitivities (POLSKY, 2004) of temperature and precipitation are examined.

Ricardian Climate Sensitivities

The Ricardian climate sensitivities are evaluated for each seasonal and annual variable and are used to assess the extent of the climate effects on land values. In the standard Ricardian model climate is modeled using a quadratic formulation, whereas the other variables are linear in the function. Thus recalling equation 7 and differentiating it with respect to a climate component f_i (e.g. summer sowing temperature) yields the following marginal value or climate sensitivity for each climate component:

$$MCI = \frac{dLV}{df_i} = \beta 1 + 2 * \beta 2 * f_i \qquad (12)$$

With MCI denoting Marginal Climate Impact, $\beta 1$ representing the linear coefficient of the climate component and $\beta 2$ denoting the squared coefficient and f_i representing the level of the climate components' long term normal value. These marginal values of each climate variable depend on the linear and quadratic coefficients. The respective climate component's marginal impact is computed by replacing E $[f_i]$ for f_i and estimating the mean of the climate component; in this regard the climate variable's level of measurement also matters. The marginal assessment reveals how a change of 1°C in temperature or 1 mm in rainfall translates into a change in land values per acre or hectare, not only annually but also seasonally and for different farm types. Additionally, to get an even more comprehensive interpretation of the regression output, Ricardian elasticities are calculated for the climate components, this is achieved by using the following equation (BENHIN, 2008):

$$\varepsilon = MCI * \frac{\bar{f}_i}{\overline{LV}} \qquad (13)$$

With ε standing for the respective climate variable's elasticity, MCI representing the Marginal Climate Impact, $\bar{f}_i$ and $\overline{LV}$ denoting the mean of the climate component and the mean of the land value variable, respectively. The elasticity indicates i.e. how a land value responds to a 1 % increase in temperature or precipitation. To verify the significance level of the Marginal Climate Impacts, an F-Test is run (AMIRASLANY, 2010).

Tab. 5: *Ricardian Climate Sensitivities (Marginal Climate Change Impacts)*

	Temperature				Precipitation		
	Total Farms	Rainfed Farms	Irrigated Farms		Total Farms	Rainfed Farms	Irrigated Farms
Summer Sowing	1.92%	-6.30%	4.04%	Spring Precipitation	-0.04%	0.71%	0.08%
Summer Growing	-11.00%	-7.69%	-16.80%	Monsoon Precipitation	-1.46%	-1.77%	-1.03%
Summer Farming Season (Kharif)	-9.08%	-13.99%	-12.76%	Summer Farming Season (Kharif)	-1.50%	-1.06%	-0.95%
Winter Sowing	24.10%	30.00%	15.80%				
Winter Growing	-21.60%	-23.90%	-10.30%	Winter Farming Season (Rabi)	1.46%	-0.26%	1.77%
Winter Farming Season (Rabi)	2.50%	6.10%	5.50%				
Annual	-6.58%	-7.89%	-7.26%	Annual	-0.04%	-1.32%	0.82%
Annual ($)	-$123.87	-$115.28	-$167.43	Annual ($)	-$0.81	-$19.33	$18.92
Annual Elasticity	-1.51	-1.79	-1.67	Annual Elasticity	-0.01	-0.82	0.19

Source: AHMED, 2013, p. 188

Temperature Impacts

The MCI analysis reveals some interesting results. For the regression of all farms, an increase in temperature by 1°C is expected to have a positive impact on land values in the summer sowing period, which compares to the spring season. The same relationship is found for the winter sowing period and land values, indicating that higher temperatures in the initial growth period of crops may be beneficial. However, the results for the respective farming season's growing period are consistently negative throughout all three models, showing that although the initial crop growth reacts positively to higher temperature, in later stages however, especially for the summer season where temperatures are at their annual peak, temperature increases would be clearly damaging to agriculture. Furthermore, higher temperatures in the growing period can be expected to increase the water demand. The overall annual marginal climate impact for the full sample reveals an annual loss in agricultural land values to the tune of -6.58% from a 1°C increase in temperature. Interestingly, when comparing rainfed and irrigated farms the results as expected reveal contrasting results for the different farm types. Although the model for all farms revealed a positive impact from a 1°C increase in temperature for the summer sowing period, the effects on rainfed or in other words dryland farms are different from the impacts on irrigated farms, a result also found in earlier studies by AHMED and SCHMITZ (2011a). Rainfed farms are expected to suffer from any increase in temperature in the summer farming season. The results for the irrigated farms are consistent with the general results obtained for the mixed farm sample (total farms). While different impacts are observed for the summer season, for the winter farming season however, they are consistent over all models, indicating a positive impact from a temperature increase for the early sowing period and negative effects from a marginal warming in the growing period of the winter farming season.

As the individual sowing and growing periods present fractions of the full farming seasons, the respective agricultural farming season's Ricardian climate sensitivity is obtained by totaling the sowing and growing period's values. For the full farms sample this implies, that a marginal increase in temperature in the summer farming season is expected to reduce land values (or capitalized crop net revenues) by -9.08%. For dryland farms the damages from a 1°C in summer farming season temperature is estimated to incur losses to farmland values to the tune of -13.99%, whereas the damages for irrigated farms amount to -12.76%. Although irrigated farms also show a very high vulnerability to warming, they seem to be slightly less bad off due to the fact that additional irrigation provides an option to farmers. Nevertheless, rainfed and irrigated farms likewise, both are expected to suffer great losses incurred by higher temperatures in the summer farming season. Warming in the winter growing season is found to have a positive impact on land values in Pakistan, with a 1°C temperature increase translating into gains of 6.1% and 5.5%, for rainfed farmland values and irrigated farmland values, respectively. From a farming season point of view the results confirm the hill-shaped trend for temperature.

The sowing period marginals are generally positive and significant, indicating that a longer season is beneficial. However, the winter growing and summer growing temperature marginals are negative. As SCHLENKER et al., (2007) have shown, hot summers put additional stress on crops, whereas cold winters help to keep back pests. Clearly, the annual impact of a marginal temperature increase is found to be negative over all farm types, with dryland farms being hit most severely. For the regressions of the mixed sample, a temperature increase by 1°C is found to decrease land values on average by -6.58% per hectare. Over all seasons, rainfed farms are to suffer losses in land values to the tune of -7.89% and irrigated farms to the tune of -7.26%.

Precipitation Impacts

The precipitation impacts are clearly more consistent over all samples, suggesting that the country's arid nature is very much dependent on precipitation throughout different farming systems. The majority of the relationships between precipitation and land values are U-shaped. This finding is essential to understand the discussion of the marginal impacts regarding a 1 mm increase in precipitation. Both farm types, rainfed and irrigated, show a positive response to an addition in accumulated spring precipitation. This shows that sufficient precipitation ahead of the actual sowing period is of paramount importance to agriculture, although the impact is relatively small with 0.71% increase in land values for rainfed farms and 0.0805% for irrigated farms. The marginal precipitation effect in the summer growing period which is the period where the bulk of the country's annual precipitation is received, is found to be negative with -1.77% loss in rainfed farmland value and -1.03% in irrigated farmland value. Over the entire summer farming season this implies that losses in land values should be expected from a 1 mm in precipitation in the summer months, with rainfed farmland values depleting by -1.06% and irrigated farmland values by -0.95%. However, it has to be kept in mind that the relationship between monsoon season precipitation and land values, as reported by the parameter estimates of the Ricardian regressions, is U-shaped. The negative values thus do not indicate a negative impact of precipitation on land values as such, they explain that a 1 mm increase is simply not sufficient and thus is to be understood as an indicator for a precipitation deficiency in the hot summer months. For the winter season the results are similar, however the irrigated farms show a positive response to a 1 mm increase with an increment in land values of 1.77%, whereas rainfed farms disclose a negative impact with -0.263%. With a U-shaped relationship calculated for the winter season as well, this goes to show that the rainfed farms do not receive the minimum amount of precipitation that they require, whereas irrigated farms have already surpassed this minimum and hence are benefitting from any further increase. Over the farming seasons the precipitation deficiency is evident in both farming systems; however more so in rainfed systems, which solely depend on rainfall for farming. Mixed results are revealed for the winter farming season, with irrigated farms being positively impacted by a marginal precipitation increase and rainfed farms being negatively hit, however again with the U-shaped relationship in mind, the negative impact revealing a precipitation deficiency. Annually, the rainfed systems show a negative impact, driven by the significant precipitation deficiency in the summer growing season. Irrigated systems show a positive trend. Irrigated farms rely heavily on groundwater and precipitation is essential in this regard, therefore considering the paramount importance of irrigated farming to the country, additional water availability will be beneficial.

The estimated annual elasticity indicates that a 1% increase in temperature will lead to 1.51% decrease in land values for the whole of Pakistan, 1.79% for rainfed farms and 1.67% for irrigated farms. As the majority of the seasonal temperature coefficients show a hill-shaped trend. The elasticities calculated for precipitation indicate that for the country as whole and for rainfed farms, annual net losses to the tune of -0.01% and -0.82% are expected with a 1% increase in precipitation, whereas for irrigated farms annual gains of 0.19% are revealed.

Spatial Impacts – Agro-Ecological Zones and Provinces

In following, the goal is to assess the marginal impacts of climate on the different Agro-Ecological Zones and provinces of Pakistan. For this purpose, the respective climatic normals are used. Using the parameter estimates of the total farms model, table 6 presents the marginal impacts of climate at the Agro-Ecological Zone level and table 7 at the provincial level. The majority of Agro-Ecological Zones suffer from additional warming in the Kharif season, foremost the northern dry mountains, the wet mountains and western dry mountains zones. In contrast, marginal benefits are found for southern irrigated plain and the Sulaiman Piedmont. The winter farming season Rabi benefits from an additional warming in the zones Barani, northern dry mountains, northern irrigated Plains, western dry mountains and the wet mountains zone. As expected the zones most severely hit also have the lowest land values per hectare, precisely under 500$.

The dry and hot regions of Sindh and Balochistan comprising the zones Dry Western Plateau, Indus Delta and the Sulaiman Piedmont will suffer losses in land values from a marginal increase in normal temperature to the tune of -15%, -27.76% and -17.1%, respectively. The intersecting zone Southern irrigated Plains/ Indus Delta will likewise suffer a loss as high as -20.29%. The Northern Mountains zone is expected to slightly benefit from an additional warming, with benefits ranging between 0.7% and 1.18%.

On the province level the highest monetary reduction is encountered by the Sindh province with -110$ per hectare with an increase in temperature and precipitation by one unit each, the first being a hill shaped relationship and the second being a U-shaped relationship. In the Punjab province from an absolute point of view, decreases around -90$/ha are found, followed by -62$ for Balochistan and -37 $/ha for the N.W.F.P. province. In the winter farming season except for the Sindh province all other provinces experience positive impacts from warming. Moreover, except for the Sindh province in all other provinces the summer farming season impact is stronger and clearly over-compensates the positive impact in the winter farming months. As far as temperature is concerned annually, all provinces experience losses in land value per acre when annual normal temperatures increase by 1°C. The losses range from -2% in N.W.F.P. and -4% in the Punjab province to -7% in Balochistan and -14% in the Sindh province.

Tab. 6: *Ricardian Climate Sensitivities on the Agro-Ecological-Zone level*

AEZ	Land Value ($/ha)	Temperature						Annual	Annual ($)
		Summer Sowing	Summer Growing	Winter Sowing	Winter Growing	Summer Farming Season (Kharif)	Winter Farming Season (Rabi)		
Barani	3543	1.1%	-10.2%	25.1%	-18.0%	-9.1%	7.1%	-2.0%	-$72
Barani/Northern Irrigated Plain	2971	-4.6%	-8.2%	25.5%	-14.5%	-12.8%	11.0%	-1.8%	-$54
Barani/Sulaiman Piedmont	844	7.0%	-12.5%	23.1%	-24.4%	-5.5%	-1.2%	-6.7%	-$57
Barani/Western Dry Mountains	1720	-0.6%	-10.6%	24.7%	-18.5%	-11.2%	6.2%	-5.0%	-$85
Dry Western Plateau	410	5.0%	-11.0%	22.0%	-31.0%	-6.0%	-9.1%	-15.1%	-$62
Indus Delta	476	6.6%	-10.4%	18.7%	-42.7%	-3.8%	-24.0%	-27.8%	-$132
Northern Dry Mountains	3792	-30.7%	-5.5%	30.6%	6.3%	-36.2%	36.9%	0.7%	$27
Northern Dry Mountains/Northern Irrigated Plain	6556	-10.2%	-9.3%	26.8%	-12.5%	-19.4%	14.3%	-5.1%	-$337
Northern Irrigated Plain	2796	5.2%	-11.9%	23.7%	-22.2%	-6.7%	1.5%	-5.1%	-$144
Northern Irrigated Plain/Western Dry Mountains	4842	-8.9%	-10.8%	25.8%	-13.0%	-19.7%	12.7%	-7.0%	-$339
Sandy Desert	1024	7.0%	-12.5%	23.1%	-24.4%	-5.5%	-1.2%	-6.7%	-$69
Sandy Desert/Northern Irrigated Plain	1474	10.9%	-12.7%	22.6%	-26.4%	-1.8%	-3.8%	-5.5%	-$82
Sandy Desert/Southern Irrigated Plain	896	19.5%	-12.8%	20.6%	-35.6%	6.7%	-15.0%	-8.3%	-$75
Southern Irrigated Plain	804	19.5%	-13.4%	19.7%	-37.7%	6.2%	-18.0%	-11.8%	-$95
Southern Irrigated Plain/Dry Western Plateau	692	20.2%	-13.1%	19.6%	-39.0%	7.1%	-19.4%	-12.2%	-$85
Southern Irrigated Plain/Indus Delta	490	17.8%	-11.9%	18.1%	-44.4%	6.0%	-26.3%	-20.3%	-$99
Sulaiman Piedmont	756	22.6%	-14.0%	17.9%	-43.6%	8.6%	-25.7%	-17.1%	-$129
Sulaiman Piedmont/Southern Irrigated plain	691	21.8%	-14.0%	20.4%	-35.7%	7.8%	-15.3%	-7.5%	-$52
Western Dry Mountains	1424	-18.5%	-8.3%	29.1%	-4.1%	-26.7%	24.9%	-1.8%	-$26
Western Dry Mountains/Dry Western Plateau	712	-23.7%	-10.7%	30.8%	-1.4%	-34.4%	29.4%	-5.0%	-$35
Western Dry Mountains/Sulaiman	693	14.9%	-13.8%	22.6%	-28.7%	1.1%	-6.1%	-5.0%	-$35
Wet Mountains	2646	-32.4%	-6.8%	30.2%	2.1%	-39.2%	32.4%	-6.8%	-$180
Wet Mountains/Northern Dry Mountains	5156	-22.7%	-5.8%	29.5%	0.1%	-28.4%	29.6%	1.2%	$61
Wet Mountains/Northern Irrigated Plain	3946	-21.3%	-6.8%	28.5%	-5.2%	-28.1%	23.3%	-4.7%	-$186

Climate Change Impacts on Agriculture and the Relevance of Adaption

continued......	Precipitation							
AEZ	Spring Precipitation	Monsoon Precipitation	Winter Precipitation	Summer Farming Season (Kharif)	Annual	Annual ($)	Total MCI (%)	Total MCI ($)
Barani	-0.2%	-1.0%	1.9%	-1.2%	0.6%	$22.7	*-1.4%*	*-$49*
Barani/Northern Irrigated Plain	-0.2%	-1.1%	1.8%	-1.3%	0.5%	$15.7	*-1.3%*	*-$39*
Barani/Sulaiman Piedmont	0.0%	-1.6%	1.2%	-1.6%	-0.4%	-$3.2	*-7.1%*	*-$60*
Barani/Western Dry Mountains	-0.3%	-1.3%	1.8%	-1.6%	0.3%	$4.9	*-4.7%*	*-$81*
Dry Western Plateau	0.2%	-1.7%	1.1%	-1.5%	-0.4%	-$1.7	*-15.5%*	*-$64*
Indus Delta	0.3%	-1.5%	0.8%	-1.2%	-0.4%	-$1.9	*-28.2%*	*-$134*
Northern Dry Mountains	-1.7%	-1.3%	3.9%	-3.0%	0.9%	$35.6	*1.6%*	*$62*
Northern Dry Mountains/Northern Irrigated Plain	-0.8%	-1.1%	3.4%	-1.9%	1.5%	$95.8	*-3.7%*	*-$241*
Northern Irrigated Plain	0.0%	-1.4%	1.3%	-1.3%	0.0%	-$.1	*-5.1%*	*-$144*
Northern Irrigated Plain/Western Dry Mountains	-0.4%	-1.4%	2.1%	-1.8%	0.3%	$15.1	*-6.7%*	*-$324*
Sandy Desert	0.2%	-1.6%	1.0%	-1.4%	-0.4%	-$4.4	*-7.2%*	*-$73*
Sandy Desert/Northern Irrigated Plain	0.1%	-1.5%	1.0%	-1.4%	-0.4%	-$5.5	*-5.9%*	*-$87*
Sandy Desert/Southern Irrigated Plain	0.3%	-1.6%	0.8%	-1.3%	-0.5%	-$4.9	*-8.9%*	*-$79*
Southern Irrigated Plain	0.3%	-1.6%	0.8%	-1.3%	-0.6%	-$4.5	*-12.4%*	*-$100*
Southern Irrigated Plain/Dry Western Plateau	0.3%	-1.6%	0.7%	-1.3%	-0.5%	-$3.7	*-12.8%*	*-$88*
Southern Irrigated Plain/Indus Delta	0.3%	-1.5%	0.7%	-1.2%	-0.5%	-$2.4	*-20.8%*	*-$102*
Sulaiman Piedmont	0.2%	-1.8%	1.1%	-1.6%	-0.5%	-$3.5	*-17.6%*	*-$133*
Sulaiman Piedmont/Southern Irrigated plain	0.3%	-1.7%	0.8%	-1.4%	-0.6%	-$4.0	*-8.1%*	*-$56*
Western Dry Mountains	-0.1%	-1.6%	1.6%	-1.7%	-0.1%	-$1.4	*-1.9%*	*-$27*
Western Dry Mountains/Dry Western Plateau	0.2%	-1.8%	1.4%	-1.6%	-0.2%	-$1.6	*-5.2%*	*-$37*
Western Dry Mountains/Sulaiman	0.3%	-1.7%	0.9%	-1.4%	-0.5%	-$3.8	*-5.6%*	*-$38*
Wet Mountains	-1.2%	-0.6%	3.4%	-1.9%	1.5%	$40.8	*-5.3%*	*-$139*
Wet Mountains/Northern Dry Mountains	-1.1%	-0.9%	3.4%	-2.0%	1.5%	$75.8	*2.7%*	*$137*
Wet Mountains/Northern Irrigated Plain	-1.2%	-0.8%	3.9%	-2.0%	1.9%	$74.8	*-2.8%*	*-$112*

Source: AHMED, 2013, p. 194

Tab. 7: Climate Sensitivities on the Province Level

Province	Land Value ($/ha)	Temperature							Precipitation					Total MCI (%)	Total MCI ($)
		Summer Sowing	Summer Growing	Winter Sowing	Winter Growing	Summer Farming Season	Winter Farming Season	Annual	Spring	Monsoon	Winter	Summer Farming Season	Annual		
Punjab	2261	7%	-12%	24%	-23%	-5%	1%	-4%	0.04%	-1.33%	1.29%	-1.29%	0.00%	*-4%*	*-$90*
Sindh	733	18%	-13%	20%	-39%	5%	-19%	-14%	0.31%	-1.59%	0.76%	-1.29%	-0.53%	*-15%*	*-$106*
N.W.F.P	3666	-11%	-8%	27%	-10%	-19%	17%	-2%	-0.68%	-1.25%	2.59%	-1.93%	0.66%	*-1%*	*-$49*
Balochistan	771	-6%	-11%	26%	-16%	-17%	10%	-7%	0.14%	-1.71%	1.24%	-1.57%	-0.32%	*-7%*	*-$56*

Source: AHMED, 2013, p. 200

For precipitation very interesting findings are revealed. As figure 5 shows the country's Southern parts comprising the Sindh and Balochistan provinces are particularly dry.

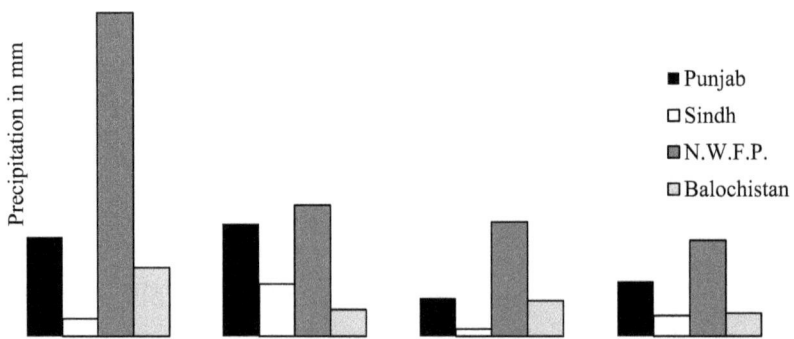

Fig. 5: Province wise long-term precipitation averages by season
Source: AHMED, 2013, p. 203

Consistent with the climatic records, the results show positive impacts from additional precipitation in the provinces Punjab and Balochistan. The results reveal the highest spring season benefits from additional accumulated rainfall for the driest province Sindh, followed by the second driest province of Balochistan and ultimately the Punjab province. Benefits range from an additional 0.305% in land value per hectare in the Sindh province and 0.139% in the Balochistan province to merely 0.041% in the Punjab province. The wettest spring is found in the N.W.F.P. (North-West Frontier Province); accordingly the results reveal a negative impact from additional precipitation in this period to the tune of -0.684%. Once again, also on the province scale, the results of the marginal analysis based on a U-shaped relationship, indicate a clear precipitation deficiency for all provinces. Every increase in precipitation although being absolutely beneficial, will be marginally negative up and until the minimum required levels of precipitation have been achieved.

Control Variables

It is important to account for the non-climate features that determine the value of land in particular geographic regions, for this different controls are necessary as no Ricardian model automatically adapts to a country's socio-economic, hydrological and pedological conditions.

To capture the urbanization alongside the effects of land use competition (demand side of hedonic model) the population density per km^2 is used as a control variable in the regression model. In line with expectations, the population density variable is positive and highly significant for the two farm typs rainfed and irrigated. Other Ricardian studies have used per capita income as a control variable for the demand side of the hedonic pricing model. The Human Development Index (HDI) offers a much more sophisticated way of measuring the level of human development in a country, serving as an indicator of both social and economic development. The interesting additional advantage of the HDI over simple income as a demand side indicator is that the HDI also indicates the policy choices of the district (in the modeled case) to some extent. In line with the proposed theory and hypothesis, the impact of human development on land values per hectare in Pakistan is significantly positive over all farm types; significant at the 99% percent level for irrigated and mixed farms and at the 95% for rainfed farms. Higher elevations in general for the whole country are found to be negatively associated with land values. A similar result is found for the rainfed farms. The irrigated farms on the other hand seem to benefit from higher elevations, also reasoned in the fact that the country's northern parts (particularly comprising Punjab and N.W.F.P.) have excellent agro-climatic foundations for successful agricultural production, these benefits translate into higher net revenues that are capitalized in land values. A proxy variable for technological abundance is also introduced. The sign on the tractor density per hectare of farmland coefficient is positive and the regressor is highly significant for all three samples. According to MALIK (2010) in rainfed (barani) areas technology is valued as a symbol of high economic status. This explains the significant positive impact of tractor density per hectare on land values, foremost in rainfed areas. A higher road density per km^2 is found to have a significant positive impact on land values per hectare for all farming systems. The model also includes a set of additional hydrological control variables, simply to the background that additional water availability will positively impact land values. Two variables have been included to control for the effect of surface water availability on land values, namely annual mean runoff and river density km^2. In line with the theoretical expectations both variables show a positive and significant impact for agriculture, foremost annual mean runoff. The runoff variable is positive and highly significant for the total farm sample and the rainfed farming systems and significant for irrigated farms. River density plays positive significant role in determining land values in Pakistan as a whole and rainfed systems. Access to surface water is thus found to give more options to farmers as they gain moderate independence from the temporal distribution and sequence of rainfall, particularly in rainfed areas. As Pakistan is a country which is frequently struck by extreme events, particularly floods, the model includes a control for flood risk, measured as the long term (1951-2010) percentage area affected by flood in each district. As the results clearly indicate, the impact of flood on land values is clearly negative over all farming systems, however significant only for the total farms' and rainfed farms' samples. To see whether interactions between the climatic and hydrological components of the model exist two types of interaction terms were introduced, one between annual mean runoff and annual temperature and one between irrigated percentage in district and temperature (only for irrigated farms). Interestingly the results reveal a contrasting picture; the runoff-temperature interaction is consistently negative over all models and significant for the mixed and rainfed sample. Contrary to the expectations in Pakistan more surface water availability does not seem to compensate for higher temperatures. The annual mean runoff does not seem to be sufficient to offset the impacts from a temperature increase, especially to the background of notably high PET rates which are close to 100%. For the irrigated farming sample an interaction term between irrigated acreage and temperature was introduced, which is found to be positive and significant. The results indicate that additional irrigation, provided water supplies are available, can compensate for higher temperatures. The control variable 'latitude' was introduced to account for solar radiation. The negative sign of the coefficient can be explained by relating latitude and solar radiation scientifically. Owing to the near spherical shape of the earth the parallel beams of solar radiation that reach the earth have a more direct impact on surfaces in lower latitudes than at higher latitudes, because at higher latitudes radiation spreads over a wider area and thus is less intense per unit surface area than at lower latitudes (NSIDC, 2012). Another important feature of Ricardian analysis is application of a control for soil differences. Work that does not include soil control variables and is restricted to one province only for studying climate change impacts (such as HANIF et. al, 2010), suffers from overestimation bias in favor of the climate variables. For these reasons, soil control variables covering the country's major soil types were introduced. The soil classes considered comprise: Calcisols, Fluvisols, Rock Outcrop and Regosols. The coefficients for the soil types show a significant impact on land values.

Simulations

By utilizing the estimated response functions the simulations incorporate the expected changes in climate delivered by an ensemble of Global Circulation Models on a district scale and calculate possible future impacts for the agriculture in the country. Country level climate change scenarios depicting changes for a ten year average for 2030, 2060 and 2090 under the SRES A2 scenario family are used (McSWEENEY et al., 2010). A Geographic Information System (GIS) is used for adding the predicted temperature change in °C for 2030, 2060 and 2090 to the baseline temperature in each district. In the baseline scenario, the farmers are assumed to continue making the same choices that they make today, provided the climate does not change. This implies that other reasons that might inflict a change in the farmer's choice behaviors over the next eighty years, are not modeled. Furthermore, the percentage change in precipitation that is predicted by the AOGCM's summarized in the UNDP climate change country profiles, is multiplied by the baseline precipitation values in each district. This procedure ensures that a new climate prediction is available for every modeled district in Pakistan. Table 8 shows seasonal and annual climate change scenarios. All values are percentage deviations from the mean climate observed between 1970 and 1999. Seasonally, precipitation is expected to decrease in the months from October to December by almost 5% and with minus 15% even more so between January and March. Spring and summer precipitation shows an increasing trend. For the 2060s the changes in precipitation for the fall and winter season remain similar to the 2030 trends, however the increases in spring and summer precipitation are much stronger, with +6.67% and +17.2% , respectively. For the long term-scenario, on a country level, table 8 shows a drastic decline in precipitation in the winter months. The fall period does not show decreases in precipitation anymore, increases are found for the three season spring, summer and fall. Using the specified climate change scenarios the original dataset was amended on the district level. The maps in figure 6 show the spatial patterns of the projected climate scenarios' impacts on land values across the country for three different farm types, mixed, rainfed and irrigated. The rainfed and irrigated farm samples do not sum up to the mixed sample. The results of the simulations clearly reveal that the country is highly vulnerable to climate change if the change is of a uniform nature, specifically rainfed farms are severely affected, with declines in land values and thus capitalized net revenues between -30% and -10% in the southern most parts of the country in the Sindh province.

Tab. 8: Pakistani average annual and seasonal Climate Change scenarios 2030-2090

	Obs	Temperature (°C)	Precipitation (%)
2030			
Annual	2369	1.521992	4.89%
JFM	2369	1.652469	-15.50%
AMJ	2369	1.644407	2.71%
JAS	2369	1.457282	3.91%
OND	2369	1.599831	-4.89%
2060			
Annual	2369	3.086661	6.91%
JFM	2369	3.14943	-15.47%
AMJ	2369	3.094344	6.67%
JAS	2369	2.952216	17.20%
OND	2369	3.089194	-10.11%
2090			
Annual	2369	5.075728	3.75%
JFM	2369	5.236809	-32.66%
AMJ	2369	5.025791	7.58%
JAS	2369	4.446813	18.86%
OND	2369	5.098227	2.78%

Note: values denote deviations from the normal climate!

Source: own calculation based on amended dataset using McSWEENEY et al., 2010

Only selected districts in the more or less humid northern most parts of the country and some parts in the western moderately humid parts reveal gains from a uniform change in climate. The SRES A2 scenarios however, show a slightly different picture.

Climate Change Impacts on Agriculture and the Relevance of Adaption   237

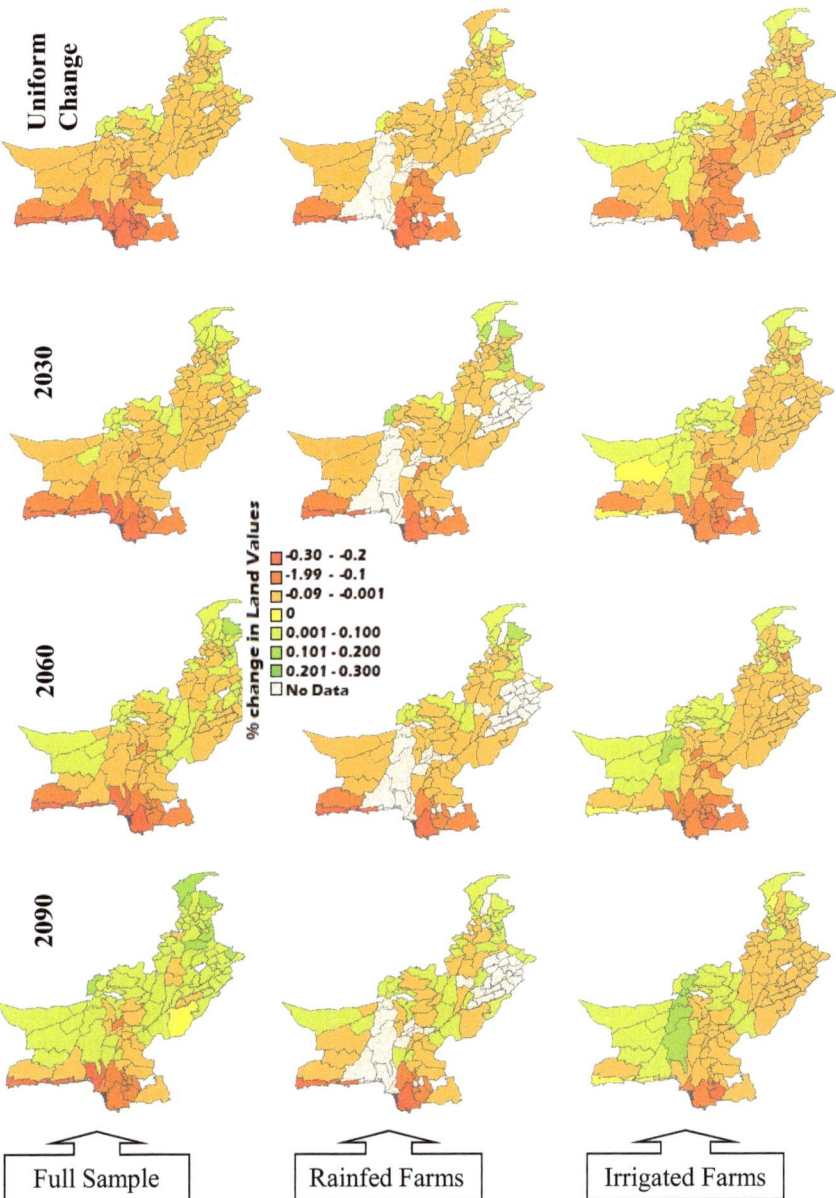

Fig. 6:   Spatial patterns of Climate Change impacts on land values – SRES A2
Source: AHMED, 2013, p. 245 (in Arc GIS 10.1 (ESRI, 2011))

With increasing temperature and increasing precipitation, the losses are reduced over the course of the three scenarios from 2030 to 2090, showing that, if farmers keep on doing what they have done so far and provided that the other things modeled remain unchanged over the next 80 years, then in most of the districts agriculture can benefit from climate change. Common in all scenarios is the significantly high negative impact from climate change in the southern parts of the country, specifically in the Indus Delta. These maps present the percentage change in land values resulting from a uniform change in climate (1°C and 1 mm – first row), a predicted change by SRES A2 ensemble 2030 (second row), ensemble 2060 (third row), and ensemble 2090 (fourth row). Further differentiation is applied with respect to farm type: first column (full sample), second (rainfed farms) and third column (irrigated). The standard Ricardian model does not model specific adaptation, but it implicitly includes adaptation, which shows that the adaptations that were made by the farmers over the course of the last 30 to 50 years have been well directed and can be expected to benefit the farmers for adaptation to future risks.

Conclusion

The economic assessment of climate change impacts poses a serious challenge as great temporal and spatial variation can be expected. This was also revealed by the present study. A major contribution is made towards the analysis of spatial effects across Agro-Ecological Zones, provinces and farm-types using a multi-seasonal approach. The models confirm the high sensitivity of farming in Pakistan to climate, the results also confirm the non-linear relationship between climate and farm income. For dryland farms the damages from a 1°C in summer farming season temperature is estimated to incur losses to farmland values to the tune of -13.99%, whereas the damages for irrigated farms amount to -12.76%. Although irrigated farms also show a very high vulnerability to warming, they seem to be slightly less bad off due to the fact that additional irrigation provides an option to farmers. Warming in the winter growing season is found to have a positive impact on land values in Pakistan, with a 1°C temperature increase translating into gains of 6.1% and 5.5%, for rainfed farmland values and irrigated farmland values, respectively. The majority of the relationships between precipitation and land values are U-shaped. The precipitation impacts are clearly more consistent over all samples, suggesting that the country's arid nature is very much dependent on precipitation throughout different farming systems. Marginally higher temperatures in the earlier parts of the respective growing season (or spring and fall) are found to be beneficial for farming. Longer sowing seasons seem to benefit farmers. The majority of Agro-Ecological Zones suffer from additional warming in the Kharif (summer) season, foremost the northern dry mountains, the wet mountains and western dry mountains zones. The winter farming season Rabi benefits from an additional warming in the zones Barani, northern dry mountains, northern irrigated plain, western dry mountains and the wet mountains zone. As expected the zones most severely hit also have the lowest land values per hectare, precisely under 500$. As far as temperature is concerned annually, all provinces experience losses in land value per acre when annual normal temperatures increase by 1°C. The losses range from -2% in N.W.F.P. and -4% in the Punjab province to -7% in Balochistan and -14% in the Sindh province. The results reveal the highest spring season benefits from additional accumulated rainfall for the driest province Sindh, followed by the second driest province of Balochistan and ultimately the Punjab province. The sensitivity of the land market to climatic extreme events is confirmed by the results for the flood risk variable. When warming is simulated further, interesting results are revealed. Winter season benefits by 2090 over-compensate summer season losses, however not in the dryer parts of the country. In general farmers in Pakistan are found to have well adapted to the current climatic conditions. Given climate change and if nothing else changes, they would not have to suffer great losses. If other factors such as water availability do change, the advantages from the beneficial effects in certain areas should be utilized, whereas the negative impacts should be limited.

List of References

Ahmed, M.N. (2013): A Structural Ricardian Valuation of Climate Change Impacts on Agriculture in Pakistan. Schriftenreihe des Zentrums für Internationale Umwelt- und Entwicklungsforschung (ZEU), 31, Peter Lang Verlag, Frankfurt a.M..
Ahmed, M.N. and Schmitz, P.M. (2011): Using the Ricardian Technique to estimate the impacts of Climate Change on Crop Farming in Pakistan- European Association of Agricultural Economists 2011 International Congress, August 30 - September 2, Zurich, Switzerland.
Ahmed, M.N. and Schmitz, P.M. (2011a): Economic Assessment of the Impact of Climate Change on the Agriculture of Pakistan. Business & Economic Horizons, BEH, 4 (1), Prague, 1-12.
Amiraslany; A. (2010): The Impact of Climate Change on Canadian Agriculture – A Ricardian Approach. PhD Thesis, University of Saskatchewan, Canada.
Barry, S., Burton, I., Klein, R. and Wandel, J. (2000): An Anatomy of Adaptation to Climate Change and Variability. Climatic Change, 45, 223-51.
Benhin, J. (2008): South African Crop Farming and Climate Change – an economic assessment. Global Environmental Change, 45, 666-678.
CAIT (2011): Climate Analysis Indicators Tool version 8.0. Washington, DC: World Resources Institute, 2011. Available at http://cait.wri.org.

CBSE (2006): Natural Hazards and Disaster Management. Supplementary Textbook in Geography. UNIT 11: Natural Hazards and Disasters. Central Board of Secondary Education, Preet Vihar, Delhi, 1-10.

Chaudhry, Q.Z., Mahmood, A., Rasul, G. and Afzaal, M. (2009): Climate Change Indicators of Pakistan, Pakistan Meteorological Department, Technical Report No. PMD. 22/2009, Islamabad.

EM-DAT (2012): The OFDA/CRED International Disaster Database, www.emdat.be - Université catholique de Louvain - Brussels - Belgium, Created on: Jan-30-2012. Data version v.12.07.

ESRI (2011): ArcGis Desktop 10.1. New York Street, Redlands, CA, USA.

Fisher, A., Hanemann, M., Roberts, M. and Schlenker, W. (2007): Potential Impacts of Climate Change on Crop Yields and Land Values in U.S. Agriculture: Negative, Significant, and Robust. University of California Energy Institute Seminar, Berkley.

Goulder, L.H., and Pizer, W.A. (2006): The Economics of Climate Change. Resources for the Future, Discussion Paper.

Hanif, U., Syed, S.H., Ahmad, R. and Malik, K.A. (2010): Economic impact of climate change on Agricultural sector of Punjab. 25th Annual Meeting of the Pakistan Institute of Development Economics.

Helms, S., Mendelsohn, R. and Neumann, J. (1996): The Impact of Climate Change on Agriculture. Climatic Change, 33, 1-6.

Houghton, J.T., Ding, D. J., Griggs, M. N., Van der Linden, P.J. and Xiaosu, D. (2001): Climate Change 2001: The Scientific Basis: Contributions of Working Group I to the Third Assessment Report of the Intergovernmental Panel on Climate Change. Cambridge University Press.

IPCC (2007): Climate Change 2007: Impacts, Adaptation and Vulnerability. Contribution of Working Group II to the Fourth Assessment Report of the Intergovernmental Panel on Climate Change, M.L. Parry, O.F. Canziani, J.P. Palutikof, P.J. van der Linden and C.E. Hanson (eds.).Cambridge University Press, Cambridge, 976.

Kaufmann, R.K. (1998): The Impact of Climate Change on US Agriculture: A Response to Mendelsohn et al. (1994). Ecological Economics, 26 (2), 113-119.

Kusters, K. and Wangdi, N. (2013): The costs of adaptation: changes in water availability and farmers responses in Punakha district, Bhutan, Int. J. Global Warming, 5 (4), 387-399.

LP (2008): LEAD Climate Change Action Program- LEAD Pakistan, Islamabad, Internal Document.

Malik, S.B. (2010): The Culture of Technology in Barani (Rainfed) Areas of Potahar: An Analysis of Farming Tools and Technologies from Gender Perspective. International Journal of Business and Social Science, 1 (3), 146-157.

McSweeney, C., New, M. and Lizcano, G. (2010): The UNDP Climate Change Country Profiles: improving the accessibility of observed and projected climate information for studies of climate change in developing countries. Bulletin of the American Meteorological Society, 91, 157-166.

Mendelsohn R., Nordhaus W. and Shaw D. (1994): The Impact of Global Warming on Agriculture: A Ricardian analysis-American Economic Review, 84, 753-771.

Mendelsohn, R. (2000): Efficient Adaptation to Climate Change. Climatic Change, 45, 583-600.

Mendelsohn, R. and Dinar, A. (2009): Climate change and agriculture: An economic analysis of global impacts, adaptation and distributional effects. Northampton, MA: Edward Elgar Publishing.

Mendelsohn, R. and Nordhaus, W. (1996): The Impact of Global Warming on Agriculture: Reply. American Economic Review, American Economic Association, 86 (5), 1312-1315.

NSIDC (2012): Arctic Climatology and Meteorology-Latitude. The National Snow and Ice Data CenterAdvancing knowledge of Earth's frozen regions, University of Colorado Boulder, CO, www.nsidc.org, accessed July 2012.

Polsky, C. (2004): Putting Space and Time in Ricardian Climate Change Impact Studies: Agriculture in the U.S. Great Plains, 1969-1992- Annals of the Association of American Geographers, 94 (3), 549-564.

Ricardo, D. (1817): On the Principles of Political Economy and Taxation. John, London.

Rivera-Ferre, M.G., Di Masso, M., Miele, M., López-i-Gelats, F., Gallar, D., Vara, I. and Cuellar, M. (2012): Understanding the Role of Local and Traditional Agricultural Knowledge in a Changing World Climate: The case of the Indo-Gangetic Plains. CGIAR Report.

Rosen, S. (1974): Hedonic prices in Implicit Markets: Product Differentiation in Pure Competition. Journal of Political Economy, 82 (1), 34-55.

Sanghi, A. (1998): Global Warming and Climate Sensitivity: Brazilian and Indian Agriculture. Ph.D. Dissertation, Dept. of Economics, University of Chicago, USA.

Schlenker, W., Hanemann, W.M. and Fisher, A.C. (2005): Will U.S. Agriculture Really Benefit From Global Warming? Accounting for Irrigation in the Hedonic Approach. American Economic Review, March 2005, 95 (1), 395-406.

Schlenker, W., Hanemann, W.M. and Fisher, A.C. (2007): Water Availability, Degree Days, and the Potential Impact of Climate Change on Irrigated Agriculture in California. Climatic Change, 81 (1), 19-38.

TFCC (2010): Task Force on Climate Change, Final Report, Planning Commission, Government of Pakistan, Islamabad.

Washington, R. (2006): African climate change: Taking the shorter route. Bulletin of the American Meteorological Society, 87, 1355-1366.

WDI (2013): World Development Indicators 2012. The World Bank, Washington D.C, USA.

Wolfe, D.W. and Erickson, J.D. (1993): Carbon dioxide effects on plants: uncertainties and implications for modeling crop response to climate change. In: Agricultural Dimensions of Global Climate Change. H.M. Kaiser and T.E. Drennen (eds.), 153-178.

Contributors

Prof em Dr Dr h.c. mult. Ulrich KOESTER
Christian-Albrechts-University Kiel, Germany
Department of Agricultural Economics

Prof Dr Monika HARTMANN
Rheinische Friedrich-Wilhelms-University Bonn, Germany
Institute for Food and Resource Economics

Dr Johannes SIMONS
Rheinische Friedrich-Wilhelms-University Bonn, Germany
Institute for Food and Resource Economics

Kakuli DUTTA
Rheinische Friedrich-Wilhelms-University Bonn, Germany
Institute for Food and Resource Economics

Prof Dr Roland HERRMANN
Justus-Liebig-University Giessen, Germany
Institute for Agricultural Policy and Market Research

Dr Rebecca SCHRÖCK
Justus-Liebig-University Giessen, Germany
Institute for Agricultural Policy and Market Research

Dr Matthias STAUDIGEL
Justus-Liebig-University Giessen, Germany
Center for international Development and Environmental Research

Dr Michaela KUHL
Bad Vilbel, Germany
Commodity Analyst

Prof Dr Christian FISCHER
Free University of Bozen-Bolzano, South Tyrol, Italy
Faculty of Science and Technology

Prof Dr Martina BROCKMEIER
University of Hohenheim, Germany
Institute for Agricultural Economics and Social Sciences in the Tropics and Subtropics

Tanja ENGELBERT
University of Hohenheim, Germany
Institute for Agricultural Economics and Social Sciences in the Tropics and Subtropics

Dr Janine PELIKAN
Federal Research Institute for Rural Areas Forestry and Fisheries, Germany
Thünen Institute of Market Analysis

Prof Dr Jong-Hwan KO
Pukyong National University Busan, South Korea
Division of International and Area Studies

Prof em Alan MATTHEWS
Trinity College Dublin, Ireland
Department of Economics

Prof Dr Khandaker M. M. RAHMAN
Bangladesh Agricultural University Mymensingh, Bangladesh
Department of Agriculture Statistics

Prof Dr Mohammad I. A. MIA
Bangladesh Agricultural University Mymensingh, Bangladesh
Department of Agribusiness and Marketing

Prof Dr Mohammad Z. ABEDIN
Bangladesh Agricultural University Mymensingh, Bangladesh
Department of Farm Structure and Environmental Engineering

Prof Dr Mohammad Z. RAHMAN
Bangladesh Agricultural University Mymensingh, Bangladesh
Department of Farm Structure and Environmental Engineering

Dr Mirza Nomman AHMED
Kleffmann Group, Lüdinghausen, Germany
AgMarket Insights Division

Bibliography Prof Dr Dr h.c. P. Michael SCHMITZ

Monographs

1. Wohlfahrtsökonomische Beurteilung preis- und währungspolitischer Interventionen auf EG-Agrarmärkten. Europäische Hochschulschriften, V, 272, Frankfurt a.M. 1980.

2. Intervention oder Markt - Absatzmöglichkeiten für konventionelle und neuartige Milchprodukte. Beiträge zur Struktur- und Konjunkturforschung, 15, Bochum 1981. (Co-authors: R. HERRMANN and D. KIRSCHKE).

3. Handelsbeschränkungen und Instabilität auf Weltagrarmärkten. Weltwirtschaftliche Studien, 21, Institut für Europäische Wirtschaftspolitik, Hamburg 1984.

4. Die zukünftige Entwicklung der Landwirtschaft in den fünf neuen Bundesländern. Kiel 1991. (Co-author: S. WIEGAND).

5. Agricultural Trade and Economic Integration in Europe and in North America. Proceedings of the 31st European Seminar of the European Association of Agricultural Economists (EAAE), Frankfurt a.M., 7-9 December 1992, Kiel 1993. (Co-authors: M. HARTMANN and H. von WITZKE).

6. Landwirtschaft und Chemie - Simulationsstudie zu den Auswirkungen einer Reduzierung des Einsatzes von Mineraldüngern und Pflanzenschutzmitteln aus ökonomischer Sicht. Kiel 1993. (Co-author: M. HARTMANN).

7. Redefining the Roles for European Agriculture. Proceedings of the 8th Congress of the European Association of Agricultural Economists. Edinburgh, Scotland, UK, 3-7 September 1996. European Review of Agricultural Economics. 24 Special Issue, 1997. (Co-author: M. HARTMANN).

8. Das EU-Agribusiness im Globalisierungs- und Transformationsprozess. In: Herrmann, R., D. Kirschke und P.M. Schmitz (eds.), Landwirtschaft in der Weltwirtschaft. Festschrift anlässlich des 60. Geburtstags von Prof. Dr. Ulrich KOESTER. Agrarwirtschaft Sonderheft 158, Frankfurt a.M. 1998, 276-304. Imprint Agribusiness-Forschung, Institut für Agribusiness, 5, Leipzig 1998.

9. Economic Transition and the Greening of Policies. Modelling New Challenges for Agriculture and Agribusiness in Europe. Proceedings of the 50th European Seminar of the European Association of Agricultural Economists (EAAE) and the Follow-Up Conference of the European Short Sourse in Global Trade Analysis, Giessen, 17-19 October 1996. Kiel 1998. (Co-authors: M. BROCKMEIER, J.F. FRANCOIS, and T.W. HERTEL).

10. Agrarpolitik und Marktforschung vor neuen Herausforderungen an der Jahrtausendwende. Giessener Schriften zur Agrar- und Ernährungswirtschaft, 27, Frankfurt a.M. 1998. (Co-authors: R. HERRMANN and E. SCHINKE).

11. Auswirkungen der Währungsunion auf die Internationale Agrarwirtschaft – Die erwarteten Wirkungen auf den Agrarhandel. Institut für Agribusiness (ed.), Agribusiness-Forschung, 7, Giessen 1998. (Co-author: M. KUHL).

12. Zur Analyse der Kosten und des Nutzens des chemischen Pflanzenschutzes in der deutschen Landwirtschaft aus gesamtwirtschaftlicher Sicht. Institut für Agribusiness (ed.), Agribusiness-Forschung,10, Giessen 1999. (Co-author: M. KISSLING).

13. Wettbewerbsfähigkeit der sächsischen Landwirtschaft. Institut für Agribusiness (ed.), Agribusiness-Forschung, 11, Giessen 1999. (Co-author: M. DILLENBURG).

14. Bewertung von Landschaftsleistungen in der Verbandsgemeinde Rennerod. Institut für Agribusiness (ed.), Agribusiness-Forschung, 13, Giessen 2000. (Co-author: M. MÜLLER).

15. Bewertung von Landschaftsleistungen in der Verbandsgemeinde Daaden. Institut für Agribusiness (ed.), Agribusiness-Forschung, 14, Giessen 2000. (Co-author: M. MÜLLER).

16. Cost-Benefit-Analysis of Crop Protection (Agrarökonomische Monographien und Sammelwerke), Wissenschaftsverlag Vauk, Kiel 2001. (Co-authors: M. KUHL and S. WIEGAND).

17. Nutzen-Kosten-Analyse Pflanzenschutz. Agricultural Economy Monographs and Collected Editions, Wissenschaftsverlag Vauk, Kiel 2002.

18. Wirtschaftliche Auswirkungen einer Kulturlandschaftsprämie. Institut für Agribusiness (ed.), Agribusiness-Forschung, 15, Giessen 2002.

19. Food Production Sustainability and Security in Bangladesh – Status and Prospects, Giessen 2007. (Co-author: K. RAHMAN).

20. Analyse und Bewertung des Milchlieferstreiks in Deutschland. Institut für Agribusiness (ed.), Agribusiness-Forschung, 19, Giessen 2008. (Co-author: J.W. HESSE).

21. Die Bedeutung nachwachsender Rohstoffe am Standort Deutschland. Institut für Agribusiness (ed.), Agribusiness-Forschung, 20, Giessen 2008.

22. Das Warengeschäft im genossenschaftlichen Verbund: Fakten, Trends und Chancen. Institut für Agribusiness (ed.), Agribusiness-Forschung, 21, Giessen 2007. (Co-authors: J.W. HESSE, S. MAAS and K. SCHMITZ).

23. Bedeutung des AgriFoodBusiness für den Standort Deutschland., Institut für Agribusiness (ed.), Agribusiness-Forschung, 22, Giessen 2008.

24. Das verfassungsrechtliche Aus des Absatzfonds – Ökonomische Bewertung und Entwurf einer Nachfolgelösung. Institut für Agribusiness (ed.), Agribusiness-Forschung, 23, Giessen 2009. (Co-author: J.W. HESSE).

25. Bedeutung des AgriFoodBusiness am Standort Deutschland. Institut für Agribusiness (ed.), Agribusiness-Forschung, 24, 3rd Edition, Giessen 2010.

26. Die Bedeutung nachwachsender Rohstoffe am Standort Deutschland., Institut für Agribusiness (ed.), Agribusiness-Forschung, 25, 2nd Edition, Giessen 2010.

27. Restricted availability of azole-based fungicides. Impacts on EU farmers and crop agriculture. Institut für Agribusiness (ed.), Agribusiness-Forschung, 27, Giessen 2011. (Co-authors: A. MATTHEWS, N. KEUDEL, S. SCHRÖDER and J.W. HESSE).

28. Agro-economic analysis of the use of Glyphosate in Germany. Institut für Agribusiness (ed.), Agribusiness-Forschung, 28, Giessen 2011. (Co-authors: M.N. AHMED, H. GARVERT and J.W. HESSE).

29. Cross Compliance und Greening – Gibt es Vorteile für landwirtschaftliche Betriebe bei Verzicht auf Direktzahlungen? Institut für Agribusiness (ed.), Agribusiness-Forschung 29, Giessen 2013. (Co-authors: J.W. HESSE and H. GARVERT).

30. Bestimmungsgründe für das Niveau und die Volatilität von Agrarrohstoffpreisen auf internationalen Märkten. Verband der Deutschen Biokraftstoffindustrie (VDB) and Union zur Förderung von Oel- und Proteinpflanzen (UFOP) (eds.), Giessen 2013. (Co-author: P. MOLEVA).

31. Bestimmungsgründe für das Niveau und die Volatilität von Agrarrohstoffreisen auf internationalen Märkten – Sind Biokraftstoffe verantwortlich für Preisschwankungen und Hunger in der Welt? Institut für Agribusiness (ed.), Imprint Agribusiness-Forschung, 30, Giessen 2013. (Co-author: P. MOLEVA).

Contributions in books and series of publications

1. Möglichkeiten der Nachfragesteigerung nach konventionellen und nichtkonventionellen Milchprodukten. Scientific Series University Kiel, 60, 1979, 183-195. (Co-authors: R. HERRMANN and D. KIRSCHKE).

2. Chancen und Risiken der Agrarwährungspolitik in einer erweiterten Gemeinschaft. In: Alvensleben, R. v., U. Koester und H. Storck (eds.), Agrarwirtschaft und Agrarpolitik in einer erweiterten Gemeinschaft. Scientific Series Gesellschaft für Wirtschafts- und Sozialwissenschaften des Landbaues e.V., 18, Münster-Hiltrup 1981, 443-464.

3. Präferenzabkommen der EG. In: Böckenhoff, E., H. Steinhauser und W.von Urff (eds.), Landwirtschaft unter veränderten Rahmenbedingungen. Scientific Series Gesellschaft für Wirtschafts- und Sozialwissenschaften des Landbaues e.V., 19, Münster-Hiltrup 1982, 405-421.

4. Auswirkungen der EG-Agrarpreispolitik auf die Instabilität landwirtschaftlicher Erlöse. In: Grosskopf, W. und M. Köhne (eds.), Einkommen in der Landwirtschaft - Entstehung, Verteilung, Verwendung und Beeinflussung. Scientific Series Gesellschaft für Wirtschafts- und Sozialwissenschaften des Landbaues e.V., 21, Münster-Hiltrup 1984, 653-671.

5. The Common Agricultural Policy and Instability on World Food Markets. In: Thomson, K.J. und R.M. Warren (eds.), Price and Market Policies in European Agriculture. Proceedings of the 6th Symposium of the European Association of Agricultural Economists, 14-16 September 1983. Newcastle upon Tyne, England 1984, 318-331.

6. European Community Trade Preferences for Sugar and Beef. In: Deutsche Forschungsgemeinschaft und Institut für wissenschaftliche Zusammenarbeit mit Entwicklungsländern (eds.), Recent German Research in International Economics - Special Research Program 86, Hamburg/Kiel, Chairman: H. Giersch, Bonn und Tübingen 1984, 106-125.

7. The Sugar Market Policy of the European Community and the Stability of World Market Prices for Sugar. In: Sarris, A.H., A. Schmitz and G.G. Storey (eds.), International Agricultural Trade - Advanced Readings in Price Formation. Market Structure, and Price Instability, Westview Press, Boulder, Colorado 1984, 235-259. (Co-author: U. KOESTER).

8. Konfliktfelder und Lösungsversuche der Europäischen Agrarpolitik. In: Hessisches Ministerium für Landwirtschaft, Forsten und Naturschutz (ed.), Scientific Series Agricultural Research Justus-Liebig-University. Lecture 15. Hochschultagung der agrar-, haushalts- und ernährungswissenschaftlichen Fachbereiche der Justus-Liebig-Universität Giessen vom 23.11.1984, XVII, Giessen 1985, 23-41.

9. Lage und Entwicklung der Agrarproduktion und der Agrarmärkte unter dem Einfluss des technischen Fortschritts in der Bundesrepublik Deutschland und der EG - Allgemeine Bestandsaufnahme und Bewertung. Schriftliche Stellungnahme zur öffentlichen Anhörung im Ausschuss für Ernährung, Landwirtschaft und Forsten des Deutschen Bundestages über „Die Zukunft der Deutschen Landwirtschaft in der EG: Risiken und Chancen im nächsten Jahrzehnt". In: Deutscher Bundestag (ed.), Ausschuss-Drucksache, 10/150 vom 21.10.1985, 15-26.

10. Internationale Auswirkungen der EG-Agrarpolitik. Schriftliche Stellungnahme zur öffentlichen Anhörung im Ausschuss für Ernährung, Landwirtschaft und Forsten des Deutschen Bundestages über „Die Zukunft der deutschen Landwirtschaft in der EG: Risiken und Chancen im nächsten Jahrzehnt". In: Deutscher Bundestag (ed.), Ausschuss-Drucksache, 10/150, 21.10.1985, 27-44.

11. Enogenous Policy Determination within the Common Agricultural Policy - Outline of a Heuristic Approach. In: International Institute for Applied Systems Analysis (ed.), Development of a Model for EC Food and Agricultural Policy Analysis - Final Report, 2, Laxenburg 1985, 1-13. (Co-author: U. FÄRBER).

12. The Impact of a Price Harmonization in the European Community and its Member Countries. In: International Institute for Applied Systems Analysis (ed.), Development of a Model for EC Food and Agricultural Policy Analysis Final-Report, 7, Laxenburg 1985, 114. (Co-authors: U. FÄRBER and P. PIERANI).

13. Opener's Comments on two invited papers of Theme 7: Food Chain, Markets and Trade. Written Comment presented at the XIX. International Conference of Agricultural Economists, Málaga (Spain), August 26 - September 4, 1985. In: Maunder, A. und U. Renborg (eds.), Agriculture in a Turbulent World Economy, Oxford 1986, 578-580.

14. Nachwachsende Rohstoffe - Zu Fragen des Binnenmarkts und der EG-Agramarktordnungen. In: Bundesministerium für Forschung und Technologie (ed.), Nachwachsende Rohstoffe - Möglichkeiten und Grenzen einer Produktion und Verwendung heimischer Pflanzen für die Industrie, Statements, 2, Bonn 1986, 89-98.

15. Neuorientierung der EG-Agrarpolitik - Beurteilung der dargestellten Grundpositionen unter den Aspekten des Welthandels, der Versorgungssicherheit und der Unabhängigkeit. In: Henrichsmeyer, W. (ed.), Neuorientierung der EG-Agrarpolitik. Agrarspectrum, Scientifis Series Dachverbandes Wissenschaftlicher Gesellschaften der Agrar-, Forst-, Ernährungs-, Veterinär- und Umweltforschung, 12, Frankfurt a.M. 1986, 61-76.

16. Solving CAP Problems with Price Policy? In: Tracy, M. und H. v. Meyer (eds.), Alternative Support Measures for Agriculture, Working Document, Maastricht 1987, 119-130.

17. Umweltwirkungen der Gemeinsamen Agrarpolitik. In: Urff, W.von (ed.), Landwirtschaft und Umwelt - Fragen und Antworten aus der Sicht der Wirtschafts- und Sozialwissenschaften des Landbaues. Scientific Series Gesellschaft für Wirtschafts- und Sozialwissenschaften des Landbaues e.V., 23, Münster-Hiltrup 1987, 375-386.

18. Agrarpolitik und pflanzliche Märkte aus landwirtschaftlicher und volkswirtschaftlicher Sicht. In: Ministerium für Ernährung, Landwirtschaft, Umwelt und Forsten in Baden-Württemberg (ed.), Informationen für die landwirtschaftliche Beratung in Baden-Württemberg, 1, Stuttgart 1987.

19. Politik für Landwirte, die Bauern bleiben wollen - Aus volkswirtschaftlicher Sicht. In: Deutsche Gesellschaft für Agrar- und Umweltpolitik e.V. (ed.), Meinungen zur Agrar- und Umweltpolitik, 14, Bonn 1987, 117-134.

20. Diskussionsbeitrag zum Thema „Politik für Landwirte, die Bauern bleiben wollen - Aus volkswirtschaftlicher Sicht". In: Deutsche Gesellschaft für Agrar- und Umweltpolitik e.V. (ed.), Meinungen zur Agrar- und Umweltpolitik, 14, Bonn 1987, 187-192.

21. Was soll mit den Menschen geschehen, die aus der Landwirtschaft ausscheiden? In: Deutsche Landwirtschaftsgesellschaft (ed.), Die Zukunft der Landwirtschaft - unternehmerische Landwirtschaft, Working Paper N/87, Frankfurt a.M. 1987, 55-73.

22. Allokations- und Verteilungswirkungen von Quotenregelungen. In: Henrichsmeyer, W. und C. Langbehn (eds.), Wirtschaftliche und soziale Auswirkungen unterschiedlicher agrarpolitischer Konzepte. Scientific Series Gesellschaft für Wirtschafts- und Sozialwissenschaften des Landbaues e.V., 24, Münster-Hiltrup 1988, 137-155. (Co-author: M. HARTMANN).

23. Beeinflussung des Ernährungsverhaltens durch staatliche Maßnahmen: Agrarpolitik - Beitrag für den Ernährungsbericht 1988 der Deutschen Gesellschaft für Ernährung, Frankfurt a.M. 1988, 182-185.

24. Neuere Entwicklungen in der Angewandten Wohlfahrtsökonomie. In: Hanf, C.-H. and W. Scheper (eds.), Neuere Forschungskonzepte und -methoden in den Wirtschafts- und Sozialwissenschaften des Landbaues. Scientific Series Gesellschaft für Wirtschafts- und Sozialwissenschaften des Landbaues e.V., 25, Münster-Hiltrup 1988, 67-77. (Co-author: M. HARTMANN).

25. Endogenous Policy Price Determination within the Common Agricultural Policy. In: Tarditi, S., K. Thomson und P. Pierani (eds.), The Effects of Reducing the Common Agricultural Protection: A Quantitative Approach, Siena 1988.

26. Bedeutung der EG Markt- und Preispolitik für die Neuorientierung der Landnutzung. In: Arbeitskreise zur Landentwicklung in Hessen der Agrarsozialen Gesellschaft (ed.), Neuorientierung der Landnutzung. Working Paper L 8 Hessisches Ministeriums für Landwirtschaft, Forsten und Naturschutz, Wiesbaden 1989, 13-19. (Co-author: E. THIESMEIER).

27. Perspektiven der pflanzlichen Produktion in der Europäischen Gemeinschaft bis zum Jahr 2000. In: Deutsche Landwirtschafts-Gesellschaft (ed.), Mit welcher Düngungsintensität in die 90er Jahre? Working Paper DLG, M 89, Frankfurt a.M. 1989.

28. Agrarpolitik in den USA. In: Bundeszentrale für Politische Bildung (ed.), Länderbericht USA 1. Scientific Series Bundeszentrale für politische Bildung, 293/I, Bonn 1990, 615-619.

29. How to Measure the Bias against Agriculture in Developing Countries? In: Campagne, P. und J. Chataigner (eds.), Producers and Consumers versus Agricultural and Food Policy in Africa, Montpellier 1990, 229-243.

30. Efficiency und Distributional Effects of EC Policy Reforms. In: Van den Noort, P.C. (ed.), Costs and Benefits of Agricultural Policies and Projects, Kiel 1990, 101-111. (Co-author: M. HARTMANN).

31. Political Economy of the Common Agricultural Policy in the European Community. Proceedings of the 9th International Conference of the Association of International Economists, Athen (Greece), Policy and Development, 3, Worcester 1991, 72-97. (Co-authors: M. HARTMANN and W. HENRICHSMEYER).

32. Land- - und Ernährungswirtschaft im europäischen Binnenmarkt und in der internationalen Arbeitsteilung. Scientific Series Gesellschaft für Wirtschafts- und Sozialwissenschaften des Landbaues e.V., 27, Münster-Hiltrup 1991. (Co-author: H. WEINDL-MAIER).

33. Die Bedeutung von Wechselkursverzerrungen in Industrie- und Entwicklungsländern für die Welternährungssituation. In: Siebke, J. (ed.), Monetäre Konfliktfelder der Weltwirtschaft, Scientific Series Verein für Sozialpolitik, 210, Berlin 1991, 549-564.

34. Impact of EC's Rebalancing Strategy on Developing Countries - The Case of Feed. In: Bellamy, M. and B. Greenshields (eds.), Issues in Agricultural Development. Sustainability and Cooperation, IAAE Occasional Paper 6, Aldershot, 1992, 51-60. (Co-author: M. HARTMANN).

35. Auswirkungen einer Liberalisierung des Weltagrarmarktes. In: Der Weltagrarhandel im Spannungsfeld ökonomischer und ökologischer Interessen, Akademie Report, Akademie für Politik und Zeitgeschehen, Vilsbiburg 1991, 41-57.

36. Auswirkungen der Extensivierung auf Agrarmärkte und Agrarstrukturen. In: DLG (ed.), Extensive Landwirtschaft - Wunschbild oder reale Chance? Vorträge der DLG Wintertagung am 17.01.1991 in Wiesbaden, Frankfurt a.M. 1991, 21-40.

37. Kurzzusammenfassung der Workshops I und II. In: Schmitz, P.M. und H. Weindlmaier (eds.), Land- und Ernährungswirtschaft im europäischen Binnenmarkt und in der interna-

tionalen Arbeitsteilung. Scientific Series Gesellschaft für Wirtschafts- und Sozialwissenschaften des Landbaues e.V., 27, Münster-Hiltrup 1991, 587-588. (Co-author: H. WEINDLMAIER).

38. Ökonometrische Analyse des griechischen und türkischen Agraraußenhandels. In: Griechenland und Türkei - Entwicklungstendenzen im Agraraußenhandel unter den Bedingungen des EG-Binnenmarktes. Branchenanalyse am Beispiel von Zitrusfrüchten, Rosinen und Tomatenmark. Giessen, Thessaloniki und Izmir 1991, 413-429. (Co-author: S. WIEGAND).

39. Free Trade versus Protectionism. In: Becker, T., R. Gray and A. Schmitz (eds.), Improving Agricultural Trade Performance under the GATT, Kiel 1992, 104-120. (Co-author: M. HARTMANN).

40. Die Bedeutung der EG-Agrarpolitik für Entwicklungsländer. In: Fachbereich Internationale Agrarentwicklung der Technischen University Berlin (eds.), Forschung für die Agrarentwicklung in der Dritten Welt, Scientific Series University Berlin, 146, Berlin 1992, 56-87.

41. Die Studie im Überblick. In: Hartmann, M. and P.M. Schmitz (eds.), Landwirtschaft und Chemie - Simulationsstudie zu den Auswirkungen einer Reduzierung des Einsatzes von Mineraldüngern und Pflanzenschutzmitteln aus ökonomischer Sicht, Kiel 1993, 1-33. (Co-author: M. HARTMANN).

42. Ökonometrische Schätzung von Eigenpreis- und Kreuzpreiselastizitäten im Produkt- und Faktorbereich der deutschen Landwirtschaft auf der Basis einer Translog-Gewinnfunktion. In: Hartmann, M. und P.M. Schmitz (eds.), Landwirtschaft und Chemie - Simulationsstudie zu den Auswirkungen einer Reduzierung des Einsatzes von Mineraldüngern und Pflanzenschutzmitteln aus ökonomischer Sicht, Kiel 1993, 169-189. (Co-author: H. DUBBERKE).

43. Sektorale und gesamtwirtschaftliche Effekte einer Reduzierung des Chemieeinsatzes und der jüngsten EG-Agrarreformen - Simulationsergebnisse auf der Basis eines numerischen allgemeinen Gleichgewichtsmodells für die Bundesrepublik Deutschland. In: Hartmann, M. und P.M. Schmitz (eds.), Landwirtschaft und Chemie - Simulationsstudie zu den Auswirkungen einer Reduzierung des Einsatzes von Mineraldünger und Pflanzenschutzmitteln aus ökonomischer Sicht, Kiel 1993, 190-228. (Co-authors: M. BROCKMEIER and J.- H. KO).

44. Introduction. In: Hartmann, M., P.M. Schmitz and H. von Witzke (eds.), Agricultural Trade and Economic Integration in Europe and in North America. Proceedings of the 31st European Seminar of the European Association of Agricultural Economists (EAAE), Frankfurt a.M., 7-9 December 1992, Kiel 1993, X-XVI. (Co-authors: M. HARTMANN and H. von WITZKE).

45. Chinas Wirtschaftsreform auf dem Lande. In: Bohnet, A. (ed.), Chinas Weg zur Marktwirtschaft - Ein Muster eines erfolgreichen Reformprogramms? Scientific Series Zentrums für regionale Entwicklungsforschung, 2 (49), Münster 1993, 363-376.

46. Die ökologisch-ökonomische Kooperation muss bezahlbar sein. In: FIP Gesellschaft zur Förderung des Integrierten Pflanzenbaues mbH (ed.), Naturnutzende Landwirtschaft. Ökologisch zweckdienlich, ökonomisch bezahlbar, pflanzenbaulich vertretbar, von der Öffentlichkeit akzeptiert, Scientific Series Integrierter Pflanzenbau, 9, Bonn 1994, 23-40.

47. Vor welchen Herausforderungen steht die Landwirtschaft - speziell der Marktfruchtbau - und wie kann sie darauf aus der Sicht der Betriebswirtschaft reagieren? In: Deutsche Landwirtschafts-Gesellschaft e.V. (ed.), Technischer Fortschritt im Spannungsfeld gesellschaftlicher Diskussion - Wie soll die Landwirtschaft mit dem technischen Fortschritt umgehen? Working Paper DLG, C/94, Frankfurt a.M. 1994, 11-34.

48. Allokations- und Verteilungswirkungen der EG-Agrarreform. In: Landwirtschaftliche Rentenbank (ed.), Verteilungswirkungen der künftigen EU-Agrarpolitik nach der Agrarreform, Scientific Series Landwirtschaftliche Rentenbank, 8, Frankfurt a.M. 1994, 257-318. (Co-authors: M. HARTMANN and M. HOFFMANN).

49. Is there a Need for Governmental Interference to Improve the Competitiveness of Rural Areas? Discussion Opening on a Plenary Paper presented by A. Sarris. In: Peters, G.H. und D.D. Hedley (eds.), Agricultural Competitiveness: Market Forces and Policy Choice, Proceedings of the 22th International Conference of Agricultural Economists, Harare, Zimbabwe 22-29 August 1994, Dartmouth 1995, 403-405.

50. Risiken und Chancen für den Agrarstandort Deutschland. In: Strothe, A. und C. Große Frie (eds.), Grenzen und Möglichkeiten für die unternehmerische Landwirtschaft in Deutschland - Handlungsmöglichkeiten unter veränderten Rahmenbedingungen, Agrarkolleg Hamburg, 1995, 18-33. (Co-author: M. MÜLLER).

51. Sicherung der Wettbewerbsfähigkeit und Schutz der natürlichen Ressourcen. In: Fördergemeinschaft Integrierter Pflanzenbau e.V. (ed.), Agrarstandort Deutschland: Ökonomische Konsequenzen intelligenter Landwirtschaft, Bonn 1995, 18-36.

52. Sicherung des Agrarstandorts Deutschland. In: Frankfurter Landwirtschaftlicher Verein e.V., 22, 1996, 61-68.

53. Die monetäre Bewertung positiver und negativer externer Effekte der Landwirtschaft - Erfahrungen und Perspektiven. In: Linckh, G. (ed.), Nachhaltige Land- und Forstwirtschaft: Expertise, Berlin 1996, 473-501. (Co-authors: A. HENZE and S. KÄMMERER).

54. The Common Agricultural Policy of the EU and its International Implications. In: Ritter, U.P. (ed.), Problems of Structural Change in the 21st Century. National and Comparative Research from Argentina, Brazil and Germany, Frankfurt a.M. 1996, 233-250. (Co-author: M. HOFFMANN).

55. Monetäre Bewertung der Kulturlandschaft in Baden-Württemberg - Bürger bewerten ihre Umwelt. In: Linckh, G. (ed.), Nachhaltige Land- und Forstwirtschaft: Expertise, Berlin 1996, 503-523. (Co-authors: S. KÄMMERER and S. WIEGAND).

56. Rechtsvorschriften und alternative umweltpolitische Maßnahmen zur Durchsetzung einer nachhaltigen Landbewirtschaftung. In: Linckh, G. (ed.), Nachhaltige Land- und Forstwirtschaft: Expertise, Berlin 1996, 599-625. (Co-author: M. MÜLLER).

57. Agrarstandort Europa im internationalen Wettbewerb. In: Kirschke, D., M. Odening und G. Schade (eds.), Agrarstrukturentwicklungen und Agrarpolitik, Scientific Series Gesellschaft für Wirtschafts- und Sozialwissenschaften des Landbaues e.V., 32, Münster-Hiltrup 1996, 11-23.

58. Wettbewerb, Osterweiterung und Ökologisierung - wie ein Wissenschaftler die europäische Agrarpolitik sieht. In: Landhandelsverband Bayern e.V. (ed.), 50 Jahre Landhandelsverband Bayern e.V., Stuttgart 1996, 47-55.

59. Landwirtschaft im 21. Jahrhundert, -"Husumer Gesprächskreis"- 31.08.-01.09.1996. In: Deutsche Gesellschaft für Agrar- und Umweltpolitik e. V. Bonn, 1996.

60. Erfahrungen aus dem Westerwald - Eine conjointanalytische Betrachtung der landwirtschaftlichen Investitionsförderung. In: Landwirtschaftliche Rentenbank (ed.), Landwirtschaftliche Investitionsförderung: Bisherige Entwicklung, aktueller Stand, Alternativen für die Zukunft, Scientific Series Landwirtschaftliche Rentenbank, 10, Frankfurt a.M. 1996, 63-109. (Co-author: M. MÜLLER).

61. CAP and Food Security. In: Rose, R., C. Tanner und M.A. Bellamy (eds.), Issues in Agricultural Competitiveness, Markets and Policies, (IAAE Occasional Paper, 7, Aldershot 1997, 157-171.

62. Agrarmärkte und Agrarpolitik heute, morgen und übermorgen. In: Akademie Deutscher Genossenschaften e.V. (ed.), Genossenschaft 2010 - Die ländliche Genossenschaft zu Beginn des nächsten Jahrhunderts, Gelbe Schlossreihe, 11, Montabaur 1997, 15-18.

63. Agrarmärkte und Agrarpolitik heute, morgen und übermorgen. In: Genossenschaftsverband Hessen/Rheinland-Pfalz/Thüringen e.V. (ed.), Landwirtschaft nach der Jahrtausendwende - Perspektiven für den Agrarstandort Ostdeutschland, Frankfurt a.M. 1997, 35-48.

64. Die deutsche Veredlungswirtschaft im europäischen und internationalen Wettbewerb - Standortbestimmung und Visionen für eine nachhaltige Entwicklung. In: Deutsche Vilomix Tierernährung GmbH (ed.), Themen zur Tierernährung 1996/97, Neuenkirchen-Vörden 1997, 65-93.

65. Landwirtschaft ohne Chemie - Eine ökonomische Betrachtung. In: Gutsche, V. (ed.), Brauchen wir den chemischen Pflanzenschutz? Mitteilungen aus der Biologischen Bundesanstalt für Land- und Forstwirtschaft, 371, Berlin, 25-30. (Co-author: T.C. WRONKA).

66. Integration der Europäischen Land- und Ernährungswirtschaft in die Weltagrarwirtschaft: Chancen und Probleme. In: Alvensleben, von R., U. Koester and C. Langbehn (eds.), Schriften der Gesellschaft für Wirtschafts- und Sozialwissenschaften des Landbaus e.V., Band 36, Tagungsband zur 40. Jahrestagung der Gesellschaft für Wirtschafts- und Sozialwissenschaften des Landbaus e.V., Wettbewerbsfähigkeit und Unternehmertum in der Land- und Ernährungswirtschaft, 4-6 October 1999, Kiel, Landwirtschaftsverlag Münster-Hiltrup, 287-303.

67. Cost-Benefit-Analysis of Chemical Crop Protection – Recent Development and Implication for Market Participants and Policy Design. In: Kuhl, M., P.M. Schmitz and S. Wie-

gand (eds.), Cost-Benefit-Analysis of Crop Protection, Agricultural Economy Monographs and Collected Editions, Wissenschaftsverlag Vauk, Kiel 2001, 2-11. (Co-author: T.C. WRONKA).

68. Ökonomische, ethische und medizinische Relevanz zur Beurteilung ausgewählter Tierhaltungsverfahren und -systeme auf der Basis der Conjoint-Analyse. In: Artgerechte Tierhaltung in der modernen Landwirtschaft – Diskussion neuer Erkenntnisse, Scientific Series Rentenbank, 17, Frankfurt a.M. 2002, 7-47. (Co-author: M. MÜLLER).

69. Pricing Environmental Services of Agriculture. Wissenschaftsverlag Vauk, Kiel 2003. (Co-authors: M. KISSLING, K. SCHMITZ and T.C. WRONKA).

70. Integrated Ecological and Economical Valuation of Land Use Systems. In: Kissling, M., K. Schmitz, P. M. Schmitz und T. C. Wronka (eds.), Pricing Environmental Services of Agriculture, Wissenschaftsverlag Vauk, Kiel 2003, 131-152. (Co-authors: K. SCHMITZ and T.C. WRONKA).

71. Wirtschaftspolitische Rahmenbedingungen einer nachhaltigen Agrar- und Ernährungswirtschaft. In: Girnau, M., L. Hövelmann, W. Wahmhoff, W. Wolf and H. Wurl (eds.), Nachhaltige Agrar- und Ernährungswirtschaft – Herausforderungen und Chancen in der Wertschöpfungskette, Deutsche Bundesstiftung Umweltschutz, 56, Göttingen 2003, 8-12.

72. Mehrgefahrenversicherungssysteme – Bestandteil einer wettbewerbsfähigen Landwirtschaft. In: Sächsisches Staatsministerium für Umwelt und Landwirtschaft (ed.), Mehrgefahrenversicherung – Chance für das Risikomanagement in der Landwirtschaft? Proceeding Internationale Grüne Woche 16 January 2003 Berlin, Dresden, 29-37.

73. Nahrungsmittelqualität und Nahrungsmittelsicherheit – Ökonomische Grundlagen. In: Bila Tserkva State Agrarian University (ed.), Food Quality and Safety in Ukraine – The Case of Meat and Milk: The Contribution of Education, Training, Business and Policy. Proceedings of an International Conference of the EU Tempus Tacis Joint European Project "Sustainable Food Chain Management", 2-5 September 2002, Bila Tserkva, Ukraine 2003, 53-55.

74. Pricing Multifunctionality by the Modelling Framework CHOICE. In: Otte, A., D. Simmering, L. Eckstein, N. Hölze and R. Waldhardt (eds.), Eco-complexity and dynamics of the cultural landscape, Proceedings of the 34th annual Conference of the Ecological Society in Giessen, 34, 2004, 319. (Co-authors: R. BORRESCH, K. SCHMITZ and T. C. WRONKA).

75. Effects of Producing or Banning G.M. Crops. – In: Evenson, R. E. und V. Santaniello (eds.), International Trade and Policies for Genetically Modified Products, CABI-Publishing, Northworthy Way, Wallingford, Oxfordshire 2004. (Co-author: J. FLATAU).

76. CHOICE – ein integriert ökonomisch-ökologisches Konzept zur Bewertung von Multifunktionalität. In: Umwelt und Produktqualität im Agrarbereich (Scientific Series Gesellschaft für Wirtschafts- und Sozialwissenschaften des Landbaus e.V.) 40, 2005. (Co-authors: R. BORRESCH, K. SCHMITZ and T. C. WWRONKA).

77. Kriterien und Rahmenbedingungen aus Sicht der Sozioökonomie. In: Frangenberg, A. (ed.), Nachhaltige Landwirtschaft – Realistische Perspektive oder ferne Vision? Scientific Series Instituts für Landwirtschaft und Umwelt (ilu) and Fördergemeinschaft Nachhaltige Landwirtschaft e.V. (FNL), 12, 2006, 55-72. (Co-author: R. BORRESCH).

78. Volks- und betriebswirtschaftliche Rahmenbedingungen. In: Frangenberg, A. (ed.), Nachhaltige Landwirtschaft: Realistische Perspektive oder ferne Vision? Scientific Series Instituts für Landwirtschaft und Umwelt (ilu) and Fördergemeinschaft Nachhaltige Landwirtschaft e.V. (FNL), 12, 2006, 87-100. (Co-author: R. BORRESCH).

79. Market and Trade Policies for Mediterranean Agriculture: The case of fruit/vegetable and olive oil. In: M'Barek, R., Wobst, P. und H.-J. Lutzeyer (eds.), Agricultural Trade – Regional Impacts in the EU, Proceedings of the Workshop on Euro-Med Association Agreements, European Commission, Directorate General, Join Research Centre, PB2006IPTS 3548, 2006, 36-47. (Co-author: A. KAVALLARI).

80. Ökonomische Auswirkungen von unterschiedlichen Produktions- und Handelsstrategien der EU beim Einsatz von gentechnisch veränderten Pflanzen. Scientific Series Gesellschaft für Wirtschafts- und Sozialwissenschaften des Landbaues e.V., 41, Münster 2006. (Co-author: J. WRONKA).

81. Economic Effects of Producing or Banning G.M. Crops. In: Evenson, R. E. und V. Santaniello (eds.): International Trade and Policies for Genetically Modified Products. CABI-Publishing, Northworthy Way, Wallingford, Oxfordshire 2006. (Co-author: J. WRONKA).

82. Wirtschafts- und Handelsbeziehungen zwischen der Türkei und den Staaten der Europäischen Union. In: Clemens, G. (ed.), Die Türkei und Europa, Hamburg 2007, 245-258. (Co-author: S. MAAS).

83. Die Auswirkung der Finanzkrise auf die deutsche Land- und Ernährungswirtschaft – eine makroökonometrische Analyse. In: Auswirkungen der Finanzkrise und volatiler Märkte auf die Agrarwirtschaft. Edmund Rehwinkel-Stiftung Rentenbank (ed.), Scientific Series, 26, 2010, 7-52. (Co-authors: M. N. AHMED, J.W. HESSE, M. KUHL and S. MAAS).

84. The Contribution of Higher Education Institutions to Development Cooperation. In: Handbook of Internationalization of European Higher Education, 6th Supplement, Berlin 2010, D 4.6. (Co-author: I. PAWLOWSKI).

85. Spekulation, agrarische Rohstoffe und Hunger – Irrtümer und notwendige Klärungen. In: DLG e.V. (ed.): Welternährung – Welche Verantwortung hat Europa? Proceeding Deutsche Landwirtschafts-Gesellschaft e.V. (DLG) Winter Conference 2012, Frankfurt a.M. 2012, 73-81. (Co-author: P. MOLEVA).

Articles in scientific journals and professional magazines

1. Alternativen der Milchmarktpolitik. Agra-Europe, Dokumentation, 21, 1979. (Co-authors: U. KOESTER and E. RYLL).

2. Der Grenzausgleich im Agrarhandel. Wirtschaftsdienst, 59, 1979, 341-345.

3. Die objektive Methode in der Agrarpolitik. Betriebswirtschaftliche Mitteilungen der Landwirtschaftskammer Schleswig-Holstein, 289, 1979, 3-8.

4. Diskussionsbeiträge: Erwiderung auf Anmerkungen von J.-V. Schrader zum Thema Grenzausgleich. Agrarwirtschaft, 28, 1979, 178-179.

5. EC Price Harmonization: A Macroeconomic Approach. European Review of Agricultural Economics, 6, 1979, 165-190.

6. Gesamtwirtschaftliche Beurteilung einer EG-Agrarpreisharmonisierung durch Abbau des Grenzausgleichs. Agrarwirtschaft, 28, 1979, 1-10.

7. Möglichkeiten der Nachfragesteigerung nach konventionellen und nichtkonventionellen Milchprodukten. Deutsche Milchwirtschaft, 31, 1980, 1228-1232. (Co-authors: R. HERRMANN and D. KIRSCHKE).

8. Zum Thema Grenzausgleich: Bemerkungen zu dem Artikel von Schmitt, G. und E. Seebohm, Einige grundsätzliche Bemerkungen zur Diskussion um den Grenzausgleich. Agrarwirtschaft, 28, 1979, 245-254. Agrarwirtschaft, 29, 1980, 11-14. (Co-author: U. KOESTER).

9. Zur Umorientierung der Milchmarktpolitik auf Molkereiebene.„Agrarwirtschaft, 29, 1980, 308-321. (Co-authors: R. HERRMANN and D. KIRSCHKE).

10. Agrarpolitische Implikationen der jüngsten Währungsbeschlüsse. Wirtschaftsdienst, 61, 1981, 615-621.

11. Bedeutung der jüngsten Währungsbeschlüsse für den Agrarsektor. Agra-Europe, Dokumentation, 22 (44), 1981.

12. Milk Policy in the European Community: Some Alternatives. Occasional Paper 12, Centre for European Agricultural Studies, Wye College University of London. Ashford, Kent 1981. (Co-authors: U. KOESTER and E. RYLL).

13. Regional Impacts of the Market Organisation for Sugar in the EC. Third Congress of European Association of Agricultural Economists, Belgrad, September 1981. Abstract in: European Review of Agricultural Economists - Special Conference Issue, 8, 1981, 340-341.

14. Zur Umorientierung der Milchmarktpolitik auf Molkereiebene: Erwiderung. Agrarwirtschaft, 30, 1981, 90-95. (Co-authors: R. HERRMANN and D. KIRSCHKE).

15. The EC Sugar Market Policy and Developing Countries. European Review of Agricultural Economics, Imprint International Food Policy Research Institute, 9, Washington 1983, 183-204. (Co-author: U. KOESTER).

16. Instability Effects of Non-Tariff Trade Barriers on World Beef Markets. Quarterly Journal of International Agriculture, 23, 1984, 115-128.

17. Der Streit um den Grenzausgleich. Wirtschaftsdienst, 64, 1984, 192-196. (Co-author: S. JÜRGENSEN).

18. Stabilizing Producers' Earnings by Fixing Agricultural prices within the EC? European Review of Agricultural Economics, 11, 1984, 395-414.

19. The International Repercussions of EC Agricultural Policy. Intereconomics, 20, 1985, 261-267.

20. Perspektiven der Agrarpolitik. Wirtschaftsdienst, 66, 1986, 12-15.

21. Dauerhaft hohe Zuwachsraten - Untersuchung über die Entwicklung der Kühlhauskapazitäten und der Nachfrage nach TK-Produkten. GV-Praxis, Zeitschrift für moderne Großverpflegung, 10, 1986, 52-59. (Co-author: T. RICKLI).

22. Agrarpolitik im Parteienstreit. Wirtschaftsdienst, 66, 1986, 233-237.

23. Dazu der Autor - Erwiderung auf einen Leserbrief zum Leitartikel 'Schafft die Agrarpolitik heute die Überschüsse von morgen'. DLG-Mitteilungen, 101 (20), 1986, 865.

24. Kurskorrektur oder Reform der Agrarpolitik - Anforderungen an eine tragfähige Lösung für die Zukunft. Deutsche Milchwirtschaft, 37 (24), 1986, 752-757.

25. Perspektiven der Agrarpolitik - Ein Plädoyer für mehr Markt. Wirtschaftsdienst, 66, 1986, 12-15.

26. Schafft die Agrarpolitik heute die Überschüsse von morgen? DLG-Mitteilungen, 101 (16), 1986, 865.

27. Nachwachsende Rohstoffe. Verbraucher Rundschau, 2-3, Bonn 1987, 35-40.

28. The Effects of the Common Agricultural Price Policy on the Third World. Quarterly Journal of International Agriculture, 26, 1987, 341-354. (Co-author: M. HARTMANN).

29. Reformvorschläge auf dem Prüfstand. Kraftfutter, 70 (6), 1987, 196-198.

30. Wie restriktiv ist die Brüsseler Agrarpreispolitik wirklich? Agra-Europe, Dokumentation, 28 (49), 1987. (Co-author: M. MÜLLER).

31. Einfluß der Agrarmarktpolitik auf Lebensmittelmärkte und Ernährungsverhalten. Jahrbuch für Absatz- und Verbrauchsforschung, 33 (4), 1987, 353-378.

32. Wie viele Landwirte zählt das Jahr 2000? DLG-Mitteilungen, 103 (24), 1988, 1273-74. (Co-author: P. ANKER).

33. Der ewige Verlierer - Dritte Welt. EG-Magazin, 1, 1988, 14-15. (Co-author: M. HARTMANN).

34. Die Handelseffekte der EG-Agrarreform. Wirtschaftsdienst, 68, 1988, 477-483. (Co-author: M. HARTMANN).

35. EC Agricultural Reform Policy - The Beginning of a New Form of Protectionism? Intereconomics, 23, 1988, 151-158. (Co-author: M. HARTMANN).

36. Abbau von Handelsverzerrungen oder neuer Agrarprotektionismus. rer.pol.-Zeitschrift für die Mitglieder der Frankfurter Wirtschaftswissenschaftlichen Gesellschaft e.V., 1989, 17-20 and Wirtschaftsdienst, 68, 1988, 477-483.

37. Entwicklung der Agrarstruktur in der BR-Deutschland bis zum Jahr 2000. Berichte über Landwirtschaft, 67 (2), 1989. (Co-author: P. ANKER).

38. Widerspruch gegen positive Bewertung des Milchquotensystems. Agra-Europe, 30 (10), 1989, 1-3. (Co-authors: U. KOESTER, A. HENZE and J.-V. SCHRADER).

39. Land- und Milchwirtschaft zwischen Ökonomie und Ökologie - Stellungnahme aus agrarökonomischer Sicht. Welt der Milch, 43 (50), 1989, 1541-1547.

40. Plädoyer für eine offene Volkswirtschaft. Anmerkungen zu dem Artikel 'Kritische Bemerkungen zum freien Welthandel mit Agrarprodukten' von G. Weinschenk. Agrarwirtschaft, 38 (12), 442-443. (Co-author: M. HARTMANN).

41. Agrarpolitik auf dem Prüfstand: Ziele nicht erreicht. Ernährungsdienst, 44 (53), 1989, 5-6. (Co-author: M. HARTMANN).

42. Grundlagen der angewandten Wohlfahrtsökonomie. Wirtschaftswissenschaftliches Studium (WiSt), 19 (7), 1990, 328-333. (Co-author: D. KIRSCHKE).

43. Mut zu einer neuen Agrarpolitik. Wirtschaftsdienst, 70 (11), 1990, 548-550.

44. Studium der Agrarökonomie: Oft stehen Sie im Wettbewerb mit den Wirtschaftswissenschaftlern, Frankfurter Allgemeine Zeitung, 107, 1988, 41. Also in: Schnorbus, A. and R. Hank (eds.), Studieren in Deutschland, Frankfurt a.M. 1990, 14-16.

45. Agriculture und Agribusiness - Adaptations to Economic and Technical Changes: Some Comments. European Review of Agricultural Economics, 18, 1991, 381-385.

46. Land- und Ernährungswirtschaft im europäischen Binnenmarkt und in der internationalen Arbeitsteilung. Berichte über Landwirtschaft, 69, 1991, 297-315. (Co-author: H. WEINDLMAIER).

47. Fast-Food im Meinungsstreit. Jahrbuch der Absatz- und Verbrauchsforschung, 37 (2), 1991, 361-367. (Co-author: M. BROCKMEIER).

48. Agrarstandort Deutschland. Agrarwirtschaft, 41 (4/5), 1992, 97-98.

49. Perspektiven für die ostdeutsche Landwirtschaft. Agra-Europe, 33 (31), 1992, 1-15. (Co-author: S. WIEGAND).

50. Das GATT-Poker - Schlechte Karten für Europas Bauern? Agronomical-Interview. Agronomical, 1, 1992, 11.

51. Agricultural Trade and Economic Integration. Journal of Economic Integration Special Issue, Iowa, 8, 1993. (Co-authors: M. HARTMANN and H. von WITZKE).

52. Den Teufelskreis durchbrechen. In: Bild der Wissenschaft, Sonderpublikation in Zusammenarbeit mit Bayer: Wege aus der Ernährungs-Krise. Neues Brot für die Welt, 7, 1996, 16-17. (Koautor: C. RIGHETTI).

53. U.S.-Farm-Bill 1996-2002. DLG-Mitteilungen, 3, 1996, 90-91. (Co-author: C. RIGHETTI).

54. Keine Extras für Landwirte. DLG-Mitteilungen, 111 (3), 1996, 108-109.

55. Sicherung des Agrarstandorts Europa. Betriebswirtschaftliche Mitteilungen der Landwirtschaftskammer Schleswig-Holstein, 493, 1996, 53-57.

56. Zu viel wettbewerbsschädliche Elemente. VDL-Journal, 45 (4), 1996, 4-5.

57. Streitgespräch: Quo vadis Agrarpolitik? Agronomical, 2, 1996, 12-15. (Co-author: F.-J. FEITER).
58. Globale marktwirtschaftliche Rahmenbedingungen der deutschen Landwirtschaft. Bauen für die Landwirtschaft - Entwicklung in der Landwirtschaft, 33 (3), 1996, 3-5.
59. Sicherung des Agrarstandorts Europa. Ländlicher Raum, 47 (6), 1996, 263-265.
60. Artgemäße und umweltgerechte Tierhaltung unter Aspekten europäischer Agrarpolitik und Agrarentwicklung. Fachbereich 17 - Agrarwissenschaften und Umweltsicherung (ed.), Agricultural Research Justus-Liebig-University Giessen, Erfurt 21 November 1996, XXIII, Giessen 1997, 121-154.
61. Introduction: Redefining the Roles for European Agriculture. European Review of Agricultural Economics, 24, 1997, VII/VIII. (Co-author: M. HARTMANN).
62. Nachhaltige Intensivierung der Landbewirtschaftung. Ein Widerspruch in sich oder die Flucht nach vorne? In: Welternährung - Welthunger. 28 October 1996 Bonn. Bundestagsfraktion Bündnis 90/Die Grünen (ed.), lang & schlüssig, 13, Bonn 1997, 17-19.
63. Wie steht es um die Wettbewerbsfähigkeit der deutschen und der europäischen Landwirtschaft im weltweiten Kontext? DLG (ed.): Die deutsche Land- und Agrarwirtschaft im internationalen Wettbewerb - Welche Beiträge kann sie zur Welternährung leisten? Working Paprer DLG, Frankfurt a.m. 1997, 6-25.
64. Economic Implications of CAP Reform and East-Enlargement of the EU. In: Recalde de Bernardi, M. L. (ed.) Structural Transformation in Latin America and Europe. Learning from Each Other's Experience. Papers and Proceedings of the Second Arnoldshain Seminar, 1997, 169-188. (Co-author: M. HOFFMANN).
65. Landwirtschaft und Makroökonomie - Abbildung ausgewählter Zusammenhänge mit einem VAR-Modell. In: Bauer, S., R. Herrmann und F. Kuhlmann (eds.), Märkte der Agrar- und Ernährungswirtschaft, Scientific Series Gesellschaft für Wirtschafts- und Sozialwissenschaften des Landbaues e.V., 33, Münster-Hiltrup 1997, 359-383 (Co-author: M. HOFFMANN).
66. Institutional and Organisational Forces Shaping the Agricultural Transformation Process: Experiences, Causes and Implications. In: Braun, J. von (ed.), Food Security, Diversification, and Resource Management: Refocusing the Role of Agriculture, Proceedings of the 23rd Conference of the International Association of Agricultural Economists, Sacramento (USA) 1997. (Co-author: C. NÖTH).
67. Die Entwicklung der Landwirtschaft aus wirtschaftlicher Sicht. In: Deutsche Gesellschaft für Agrar- und Umweltpolitik (eds.), Landwirtschaft- ein multifunktionelles Unternehmen. Die Bedeutung der Landwirtschaft für die Gesellschaft. Meinungen zur Agrar- und Umweltpolitik, 33, Bonn 1998, 15-36.
68. Die Bewertung von Kulturlandschaften mit Hilfe der Contingent-Valuation-Methode: In: Rheinischer Verein (ed.), Kulturlandschaft Rheintal, Köln 1998. (Co-authors: S. KÄMMERER and T.C. WRONKA).

69. Bewertung von Landschaftsfunktionen. In: Steinhardt, U. und M. Volk (eds.), Regionalisierung in der Landschaftsökologie, Paper Umweltforschungszentrums Leipzig, 10, 1998. (Co-authors: M. MÜLLER and H. THIELE).

70. Dynamic Linkages of Macroeconomic and Agricultural Sector Variables in Germany: In: Brockmeier, M., J.F. Francois, T.W. Hertel und P.M. Schmitz (eds.), Economic Transition and the Greening of Policies, Modeling New Challenges for Agriculture and Agribusiness in Europe. Proceedings of the 50th European Seminar of the European Association of Agricultural Economists (EAAE) and the Follow-Up Conference of the European Short Course in Global Trade Analysis, Giessen, 17-19 October 1996. Kiel 1998, 305-328. (Co-author: M. KUHL).

71. Die gemeinsame Agrarpolitik und die Osterweiterung. In: Kühnhardt, L. (ed.), Agenda 2000 - Herausforderungen an die Europäische Union und an Deutschland, Bonn 1998. (Co-author: T.C. WRONKA).

72. Die Land- und Ernährungswirtschaft Osteuropas im europäischen Integrationsprozess. In: Clement, G. (ed.), Europäische Integration, Hamburg 1998. (Co-author: T.C. WRONKA).

73. Die Chancen in der Nahrungskette nutzen. Interview. DLG-Mitteilungen, 113 (4), 1998, 14-15.

74. Reform der Europäischen Agrarpolitik im Zeichen der Agenda 2000. Zeitschrift für Wirtschaftspolitik, 2, 1998.

75. Landwirtschaft im globalen Wandel – Herausforderungen und Handlungs-bedarf. In: Boland, H. und T. Michaelis (eds.), Herausforderungen an das Management von Agrargenossenschaften – Mitarbeiterführung und Erfolgssicherung, Series Ländliches Genossenschaftswesen, 45, Giessen 1999, 71-87.

76. Die Gemeinsame Agrarpolitik und die Osterweiterung. In: Wittschork, P. (ed.), Agenda 2000 – Herausforderungen an die Europäische Union und an Deutschland. Scientific Series Zentrums für Europäische Integrationsforschung, 8, Nomos Baden-Baden 1999, 81-96. (Co-author: T.C. WRONKA).

77. Integrierte ökonomische und ökologische Bewertung von Landschaftsfunktionen. In: Steinhardt, U. und M. Volk (eds.), Regionen in der Landschaftsökologie, Scientific Series UFZ Leipzig/Halle, Stuttgart and Leipzig 1999, 360-376. (Co-authors: M. MÜLLER, M. and H. THIELE).

78. Landwirtschaft vor großen Herausforderungen. Editorial, Rheinische Bauernzeitung, Wochenblatt für das nördliche Rheinland-Pfalz und das Saarland, 51/52, 1999, S. 18.

79. Doppelte Dividende oder doppelte Belastung? Agrarwirtschaft, 48 (2), 1999, 93-94.

80. Nutzen-Kosten-Analyse des Pflanzenschutzmittel-Einsatzes. Agrarwirtschaft, 48 (7), 1999, 269-274. (Co-authors: M. KISSLING).

81. Landwirtschaft ohne Chemie - Eine ökonomische Betrachtung. In: Gutsche, V. (ed.), Brauchen wir den chemischen Pflanzenschutz? Mitteilungen aus der Biologischen Bundesanstalt für Land- und Forstwirtschaft, 371, 2000, Berlin, 25-30. (Co-author: T.C. WRONKA).

82. Macroeconomic Shocks and Trade Responsiveness in Argentinia – A VAR Analysis. In: Konjunkturpolitik – Zeitschrift für angewandte Wirtschaftsforschung, 46 (1-2), 2000, 62-92. (Co-author: M. KUHL).

83. The economic impacts of reduced agricultural chemical use in Germany. Poster im Rahmen des 3rd International Weed Science Congress, Foz do Iguassu – Brazil 2000. (Co-author: M. KISSLING).

84. Integration der Europäischen Land- und Ernährungswirtschaft in die Weltagrarwirtschaft: Chancen und Probleme. In: Wettbewerbsfähigkeit und Unternehmertum in der Land- und Ernährungswirtschaft, Scientific Series Gesellschaft für Wirtschafts- und Sozialwissenschaften des Landbaus e.V., 36, 2000, 287-303.

85. Macroeconomic Shocks and Trade Responsiveness in Argentinia – A VAR Analysis. In: Konjunkturpolitik – Zeitschrift für angewandte Wirtschaftsforschung, 46 (1-2), 2000, 62-92. (Koautorin: M. KUHL).

86. Mehr Rationalität in der Agrarwirtschaft. Wirtschaftsdienst - Zeitschrift für Wirtschaftspolitik, 81 (2), 2001, 80-82.

87. Eine agrarpolitische Wende gegen die Vernunft? Ernährungsdienst, 15, 2001.

88. Wende zum Guten? Die geplante Reform der Agrarpolitik führt nicht zu einer umweltfreundlicheren und artgerechteren Landwirtschaft. Frankfurter Allgemeine Zeitung, 119, 2001, 14.

89. Integrierte ökonomische und ökologische Bewertung der Landnutzung in Peripheren Regionen. Berichte über Landwirtschaft, 79 (1), 2001, 19-48. (Co-authors: M. MÜLLER, H. THIELE and T.C. WRONKA).

90. Cost-Benefit-Analysis of Chemical Crop Protection – Recent Development and Implications for Market Participants. In: Kuhl, M., P.M. Schmitz und S. Wiegand (eds.), Cost-Benefit-Analysis of Crop Protection, Agrarökonomische Monographien und Sammelwerke, Kiel 2001, 2-11. (Co-author: T.C. WRONKA).

91. Integrierte ökonomische und ökologische Bewertung der Landnutzung in peripheren Regionen. Berichte über Landwirtschaft, 79 (1), 2001, 19-48. (Co-authors: M. MÜLLER, H. THIELE and T.C. WRONKA).

92. Mehr Rationalität in der Agrarwirtschaft. Wirtschaftsdienst - Zeitschrift für Wirtschaftspolitik, 81 (2), 2001, 80-82.

93. Nutzen-Kosten-Analyse und agrarpolitische Implikationen des chemischen Pflanzenschutzes. In: Biologische Bundesanstalt für Land- und Forstwirtschaft unter Mitwirkung des amtlichen Pflanzenschutzdienstes der Länder (ed.), Nachrichtenblatt des Deutschen Pflanzenschutzdienstes, 11, Braunschweig 2001, 295-297.

94. Was Pflanzenschutz kostet und was er nützt. In: Neue Landwirtschaft, Sonderheft Moderner Pflanzenschutz, Deutscher Landwirtschaftsverlag, Berlin 2001, 15-19. (Co-authors: M. BROCKMEIER, J.-H. KO and M. KISSLING).

95. Halbzeitbewertung der Agenda 2000: Chancen und Risiken von Fischlers Reformkurs. In: Agrarwirtschaft, 51 (6), Frankfurt a.M. 2001, 289-291.

96. Mid-term Review of Agenda 2000: Possibilities and Risks arising from Fischler's Path of Reform. Intereconomics, 37 (5), 2002, 238-241.

97. Cost-Benefit-Analysis of Crop Protection – An Overview. Journal of Plant Diseases and Protection, Special Issue, 18, 2002, 23-41.

98. Kommt die Agrarwende oder nicht? Innovation, 4, 2002, 4-5.

99. Neuorientierung der Agrar- und Ernährungswirtschaft in der Ukraine – Modernisierung in Wissenschaft und Lehre. Spiegel der Forschung, 19 (1), 2002, 46-49. (Co-author: I. PAWLOWSKI).

100. Die ökonomische Bedeutung der Kartoffel für das Agribusiness. Kartoffelbau, 9/10, 2002, 348-353. (Co-author: T.C. WRONKA).

101. Bewertung von Landschaftsfunktionen mit Choice Experiments. Agrarwirtschaft, 52 (8), 2003, 379-389. (Co-authors: K. SCHMITZ and T.C. WRONKA).

102. Neue Impulse für altbewährte Kooperation. Uni-Forum, 5, 2003, 9. (Co-author: J. PELIKAN).

103. The real rate of protection: the income and insurance effects of agricultural policy. Applied Economics, 36 (16), 2004, 1851-1859. (Co-authors: S.R. THOMPSON, N. IWAI and B.K. GOODWIN).

104. CHOICE – ein integriert ökonomisch-ökologisches Bewertungskonzept. Beitrag zur 44. Jahrestagung der Gesellschaft für Wirtschafts- und Sozialwissenschaften des Landbaues e.V. Umwelt- und Produktqualität im Agrarbereich, Berlin, 27-29 September 2004. http://www.agrar.huberlin.de/GEWISOLA2004/dokumente/volltexte/53_w.pdf. (Co-authors: R. BORRESCH, K. SCHMITZ and T.C. WRONKA).

105. Laudatio anlässlich der Emeritierung von Prof. Dr. Dr. h.c. Ulrich KOESTER an der Christian-Albrechts-University (CAU) zu Kiel. In: Agrar- und Ernährungswissenschaftliche Fakultät der CAU, Vorträge zur Hochschultagung 2004 und zur Verabschiedung von Prof. Dr. Dr. h.c. Ulrich KOESTER und Prof. Dr. Reimar von ALVENSLEBEN. Scientific Series der Agrar- und Ernährungswirtschaftlichen Fakultät, 102, Kiel 2004, 305-311.

106. Factors Influencing Agricultural Growth in Bangladesh. Journal of Rural Development (JRD) in South Korea, 2006. (Co-authors: K.M M. RAHMAN, J.-H. KO and J. WRONKA).

107. Wettbewerbsfähigkeit der hessischen Landwirtschaft. In: Veranstaltungen des Frankfurter Landwirtschaftlichen Vereins e.V., 32, 2006. (Co-author: S. MAAS).

108. Reform der GAP und Erweiterung der EU - Auswirkungen auf die Agrarmärkte. Contributed Paper, 8, Greece Conference Agriculture Economy, Thessaloniki, 26 November 2006, CD-ROM. (Co-authors: A. KAVALLARI and R. BORRESCH).

109. Market and Trade Policies for Mediterranean Agriculture: The case of fruit/vegetable and olive oil - Trade analysis with AGRISIM. Background paper Workshops "Euro-Med Association Agreements: Agricultural Trade – Regional Impacts in the EU", EU-Kommission, Workshopsmaterialien, Brüssel 2006. (Co-author: A. KAVALLARI).

110. Modelling Agricultural Policy Reforms in the Mediterranean Basin – Adjustments of AGRISIM. Contributed Paper anlässlich des 98. EAAE-Seminars "Marketing Dynamics within the Global Trading System: New Perspectives", Chania, Griechenland, 29 Juny - 2 July 2006. (Co-authors: A. KAVALLARI and R. BORRESCH).

111. Bericht über die 46. Jahrestagung der Gesellschaft für Wirtschafts- und Sozialwissenschaften des Landbaues (GEWISOLA) 2006 in Giessen. „Bundesministerium für Ernährung, Landwirtschaft und Verbraucherschutz (ed.), Berichte über Landwirtschaft, 85 (3), 2007, 475-502. (Co-authors: R. BORRESCH, J.W. HESSE, A. KAVALLARI, F. KUHLMANN, N. MÖSER, K. SCHMITZ and J. STOLL).

112. Unidirectional external effect from shrimp farming on yield and technical efficiency of rice farming in southern Thailand. Submitted paper to the Journal Paddy and Water Environment, 2007. (Co-authors: S. KIATPATHOMCHAI and A. BABU).

113. Gemeinsame Agrarpolitik der EU. Wirtschaftsdienst, 87 (2), Hamburg 2007, 94-100 (Co-author: S. MAAS).

114. Entwicklung der Getreidemärkte weltweit. Getreide-Magazin, 12 (2), 2007, 142-145. (Co-author: S. MAAS).

115. The curse of Meaning Well. European Voice, 13 (37), 2007, 21.

116. Mengenkorsett hat ausgedient – Zeitplan für Reform der Gemeinsamen Agrarpolitik kommt zu spät. Ernährungsdienst, 92, Frankfurt a.M. 2007.

117. Investigating External Effects of Shrimp Farming on Rice Farming in Southern Thailand - A Technical Efficiency Approach. Paddy and Water Environment, 6 (3), 2008, 319-326. (Co-authors: S. KIATPATHOMCHAI, A. BABU and S. THONGRAK).

118. Internationale Nahrungsmittelkrise: Ursachen und Massnahmen. Wirtschaftsdienst, 88 (5), 2008, 286-287.

119. Crop Plants versus Energy Plants – On the International Food Crisis. Journal of Bioorganic and Medicinal Chemistry, 2008. (Co-author: A. KAVALLARI).

120. Multilateral Trade Liberalisation and Preference Erosion: Effects on the Agricultural Sector of the Mediterranean Partner Countries of the EU, 2008. (Co-author: A. KAVALLARI).

121. Analyse und Bewertung des Milchlieferstreiks in Deutschland. In: Ministerium für Ernährung und Ländlichen Raum, Baden-Württemberg (ed.), Landinfo, 7, 2008, 37-42. (Co-author: J.W. HESSE).

122. Gechlortes Geflügelfleisch? Pro und Contra Diskussion. Lebensmittel Praxis, 13, July 2008, 8.

123. Welchen Weg geht die Agrarpolitik? In: Deutsche Welthungerhilfe e.V. (ed.), Welternährung, 2008, 12-13.

124. Evidence on Euromediterranean Trade Integretion: The Case of German Olive Oil Imports. In: German Journal of Agricultural Economics. Agrarwirtschaft, 59 (1), 2010, 40-46. (Co-authors: A. KAVALLARI and S. MAAS).

125. Evaluating Today's Landscape Multifunctionality and Providing an Alternative Future: A Normative Scenario Approach. "Ecology and Society", 15 (3), 2010, 30. Online veröffentlicht unter: http://www.ecologyandsociety.org/vol15/iss3/art30/. (Co-authors: R. WALDHARDT, M. BACH, R. BORRESCH, L. BREUER, T. DIEKÖTTER, H. FREDE, S. GÄTH, O. GINZLER, T. GOTTSCHALK, S. JULICH, M. KRUMPHOLZ, F. KUHLMANN, A. OTTE, B. REGER, W. REIHER, K. SCHMITZ, P. SHERIDAN, D. SIMMERING, C. WEIST, V. WOLTERS und D. ZÖRNER).

126. Analysing Agricultural Productivity Growth in a Framework of Institutional Quality - Paper selected for the IAMO Forum „Institutions in Transition –Challenges for New Modes of Governance", Halle (Saale) - published on Ageconsearch, 16-18 June 2010. (Co-authors: M N. AHMED and S. MAAS).

127. Entwicklungsperspektiven des deutschen Agribusiness bei unterschiedlichen Politikszenarien. Bauernzeitung 44, 2010, 30.

128. Neues Leitbild und Politikkonzept gefordert. VDL-Journal, 6, 2010, 12-13.

129. Economic Assessment of the Impact of Climate Change on the Agriculture of Pakistan. Business & Economic Horizons, BEH, 4 (1), 2011, 1-12. (Co-author: M.N. AHMED).

130. Efficiency of Cash and Food Crops Producing Farms across Three Districts in Northern Tadjikistan - a Non-parametric Approach. Australian Journal of Basic and Applied Sciences, 4, 2011, 5705-5716. (Co-authors: M. GOIBOV and B.A. ALEMU).

131. Examining the Determinants of Olive Oil Demand in Nonproducing Countries: Evidence from Germany and the UK. Journal of Food Products Marketing, 17, 2011, 355-372. (Co-authors: A. KAVALLARI and S. MAAS).

132. Using the Ricardian Technique to Estimate the Impacts of Climate Change on Crop Farming in Pakistan. European Association of Agricultural Economists, International Congress, 30.08.-02.09.2011, Zürich, Schweiz. http://purl.umn.edu/114217. (Co-author: M.N. AHMED).

133. Boomende Weltmärkte – Sind Branche und Agrarpolitik richtig aufgestellt? Deutsche Molkereizeitung (DMZ), 132 (10), 2011, 3. (Co-author: J.W. HESSE).

134. Lebensmittelkennzeichnung Region „schlägt" Öko. Deutsche Molkereizeitung (DMZ), 132 (11), 2011, 28-29. (Co-author: S. SCHRÖDER).

135. Die ökonomische Bedeutung des Wirkstoffes Glyphosat für den Ackerbau in Deutschland. Journal für Kulturpflanzen, 64 (5), 2012, 145-149. (Co-author: H. GARVERT).

136. Application of a Choice Experiment to Estimate Farmer's Preferences for Different Land Use Options in Northern Tajikistan. Journal of Sustainable Development, published by Canadian Center of Science and Education, 5 (5), 2012, 2-16. (Co-authors: M. GOIBOV, S. BAUER and M.N. AHMED).

137. Agriculture without Azoles? Agriculture, 1, 2012, 24-25. (Co-authors: N. KEUDEL and J.W. HESSE).

138. Agro-Economic Analysis of the Use of Glyphosate in Germany. Outlooks on Pest Management, 24 (2), 2013, 81-85. (Co-authors: H. GARVERT and M.N. AHMED).

139. Measuring Agricultural Support for Tajikistan. Canadian Journal of Agricultural Science, 2013. (Co-authors: P. KHAKIMOV and I. PAWLOWSKI).

140. Climate Change Impacts and the Value of Adaptation – Can crop adjustment help farmers in Pakistan? International Journal of Global Warning, special issue of the UN, 2013. (Co-author: M.N. AHMED).

141. Cross Compliance in der Landwirtschaft. Deutsche Molkereizeitung (DMZ), 7, 2013, 31-33. (Co-authors: H. GARVERT and J.W. HESSE).

142. Agrarspekulation mit Indexfonds: Wie sie funktioniert. Was sie bewirkt. Policy Brief des Leibnitz-Institut für Agrarentwicklung in Mittel- und Osteuropa (IAMO), 12, 2013. (Co-authors: T. GLAUBEN, S. PREHN, I. PIES, M.G. WILL, J.-P. LOY, A. BALMANN, B. BRÜMMER, T. HECKELEI, H. HOCKMANN, D. KIRSCHKE, U. KOESTER, R. LANGHAMMER, K. SALHOFER, S. TANGERMANN, H. von WITZKE and J. WESELER).

143. Das ZEU auf Erfolgskurs – seit 15 Jahren internationale und interdisziplinäre Entwicklungs- und Umweltforschung. Spiegel der Forschung – Wissenschaftsmagazin, 30 (2), 2013, Giessen.

Discussion papers

1. Wohlfahrtstheoretische Beurteilung einer EG-Agrarpreisharmonisierung durch Abbau des Grenzausgleichs. Working Paper 40 Institut für Agrarökonomie University Göttingen, Göttingen 1978.

2. Alternativen der Milchmarktpolitik - Analyse alternativen Entscheidungen des Agrarministerrats im Frühjahr 1979. Constributions to Discussions Instituts für Agrarpolitik und Marktlehre, University Kiel, 33, Kiel 1979. (Co-authors: U. KOESTER and E. RYLL).

3. Zur Umorientierung der Milchmarktpolitik auf Molkereiebene. Constributions to Discussions Instituts für Agrarpolitik und Marktlehre, University Kiel, 36, Kiel 1980. (Co-authors: R. HERRMANN and D. KIRSCHKE).

4. Der Einfluss der EG-Zuckerpolitik auf die Entwicklungsländer. Constributions to Discussions Instituts für Agrarpolitik und Marktlehre, University Kiel, 42, Kiel 1981. (Co-author: U. KOESTER).

5. Exporterlösstabilisierung durch AKP-Präferenzabkommen. Constributions to Discussions Instituts für Agrarpolitik und Marktlehre, University Kiel, 47, Kiel 1982.

6. World Price Effects of Protective Trade Policies under Uncertainty. Constributions to Discussions Instituts für Agrarpolitik und Marktlehre, University Kiel, 51, Kiel 1983. (Co-author: D. KIRSCHKE).

7. Diskussionsbeitrag zur "Europäischen Milchmarktpolitik auf dem Prüfstand. Ein Jahr Mengenbegrenzung - Dauerlösung oder Eintagsfliege?" In: Deutsche Gesellschaft für Agrar- und Umweltpolitik (ed.), Meinungen zur Agrar- und Umweltpolitik, 12, Bonn 1985, 140-143.

8. The Sugar Market Policy of the European Community and the Stability of World Market Prices for Sugar. In: International Food Policy Research Institute, Washington 1985. (Co-author: U. KOESTER).

9. Exchange Rates und EC-Agriculture - International and European Monetary Challenges. Contributed Paper the 6th European Congress of Agricultural Economists in Den Haag (Netherlands), September 3-7, 1990, issued in Agrarökonomische Diskussionsbeiträge der Universität Giessen, 2, Giessen 1990. (Co-author: M. HARTMANN).

10. Political Economy of the Common Agricultural in the European Community. Working Paper 2, Frankfurter Volkswirtschaftliche Diskussionsbeiträge, Faculty Economics, University Frankfurt, Frankfurt a.M. 1990. (Co-authors: M. HARTMANN and W. HENRICHSMEYER).

11. Impact of EC's Rebalancing Strategy on Developing Countries - The Case of Feed. Arbeitspapier 13, Constributions to Discussions, Faculty Economics, University Frankfurt, Frankfurt a.M. 1991. (Co-author: M. HARTMANN).

12. Impact of EC's Rebalancing Strategy on Developing Countries - The Case of Feed. Staff Paper 91-18, Department of Agricultural and Applied Economics, University of Minnesota, Minnesota 1991. (Co-author: M. HARTMANN).

13. The Economic Implications of Chemical Use Restrictions in Agriculture. Arbeitspapier 56, Constributions to Discussions, Faculty Economics, University Frankfurt, Frankfurt a.M. 1994. (Co-author: M. HARTMANN).

14. Chemical Use Restrictions in German Agriculture - Simulation Results of a Computable General Equilibrium Model. Working Paperr 52, Constributions to Discussions, Faculty Economics, University Frankfurt, Frankfurt a.M. 1994. (Co-authors: M. BROCKMEIER and J.-H. KO).

15. Instrumente und Modelle zur Analyse und Formulierung von Agrarpolitiken in Entwicklungsländern. Working Paper 49, Constributions to Discussions, Faculty Economics, University Frankfurt, Frankfurt a.M. 1994. (Co-authors: M. HARTMANN and C. NÖTII).

16. The Economic Implications of Chemical Use Restrictions in Agriculture. International Agricultural Trade Research Consortium, Working Paper 94-6. St. Paul 1994.

17. Do Developed Exporting Countries Contribute to Food Security? The Case of the EC. Staff Paper 91-34, Department of Agricultural und Applied Economies, University of Minnesota, Minnesota 1991.

18. Landwirtschaft und Makroökonomie - Abbildung ausgewählter Zusammenhänge mit einem VAR-Modell. Agrarökonomische Diskussionsbeiträge des Instituts für Agrarpolitik und Marktforschung, 39, Giessen 1996 (Co-author: M. HOFFMANN).

19. Zusammenfassung und Bewertung aus agrarökonomischer und agrarpolitischer Sicht. In: Ergebnisse landwirtschaftlicher Forschung. XXIV, Giessen 1999, 71-78.

20. Experiences and current issues of economic transition in agriculture and agribusiness the case of Germany. In: EU Tempus Tacis JEP "AGFED" (ed.). Educational provision for agricultural and food economics in the EU and Ukraine: Comparative persepectives. Proceedings of the International scientific and practical conference. Bila Tserkva 1999, 14-18. (Co-author: C. NÖTH).

21. Appraising the Environmental Performance of Rice Production Systems in Southern Thailand: A Nonparametric Approach. Accepted Paper will be published in the Proceedings of the First PSU-UNS Joint Conference on Bioscience: Food, Agriculture, and the Environment, 2007. (Co-authors: S. KIATPATHOMCHAI and S. THONGRAK).

22. Supply Chain Analysis of Olive Oil in Germany. Discussion Paper 35, Zentrum für Internationale Entwicklungs- und Umweltforschung, University Giessen 2007. (Co-authors: J. FLATAU, V. HART and A. KAVALLARI).

23. Supply Chain Analysis of Fresh Fruit and Vegetables in Germany. Discussion Paper 36, Zentrum für Internationale Entwicklungs- und Umweltforschung, University Giessen 2007. (Co-authors: V. HART, A. KAVALLARI and T.C. WRONKA).

24. An empirical assessment of trade policies in the Mediterranean basin – regional effects on the EU Member States. Contributed Paper at the 1st Mediterranean Conference of Agro-Food Social Scientists – 103rd EAAE Seminar "Adding Value to the Agro-Food Supply Chain in the Future Euromediterranean Space", Barcelona, Spanien, 23.-25. April 2007. Verfügbar auf Tagungsmaterialien und auf der Internet-Datenbank „AgEcon Search" (http://ageconsearch.umn.edu/). (Co-author: A. KAVALLARI).

25. Spekulation mit agrarischen Rohstoffen verhindern. Contribution „Öffentliche Anhörung des Ausschusses für Ernährung. Landwirtschaft und Verbraucherschutz des Deutschen Bundestags, 43. Sitzung. Berlin 27 June 2011.

Collaboration on scientific opinions of the „Wissenschaftlicher Beirat beim Bundesministerium für Ernährung, Landwirtschaft und Forsten"

1. Reduzierung der Stickstoffemissionen der Landwirtschaft. Opinion of Wissenschaftlicher Beirat beim Bundesministerium für Ernährung, Landwirtschaft und Forsten (BMELF). Schriftenreihe des BMELF, A, Angewandte Wissenschaft, 423, Münster-Hiltrup 1993.

2. Vorschläge für eine grundlegende Reform der EG-Zuckermarktpolitik. Opinion of Wissenschaftlicher Beirat beim Bundesministerium für Ernährung, Landwirtschaft und Forsten (BMELF). Schriftenreihe des BMELF, A, Angewandte Wissenschaft, 430, Münster-Hiltrup 1994.

3. Agrarpolitik und Agrarstruktur. Opinion of Wissenschaftlicher Beirat beim Bundesministerium für Ernährung, Landwirtschaft und Forsten (BMELF). Schriftenreihe des BMELF, A, Angewandte Wissenschaft, 433, Münster-Hiltrup 1994.
4. Forstpolitische Rahmenbedingungen und konzeptionelle Überlegungen zur Forstpolitik. Opinion of Wissenschaftlicher Beirat beim Bundesministerium für Ernährung, Landwirtschaft und Forsten (BMELF). Schriftenreihe des BMELF, A, Angewandte Wissenschaft, 438, Münster-Hiltrup 1994.
5. Zur künftigen Gestaltung des Agrimonetären Systems in der EU. Opinion of Wissenschaftlicher Beirat beim Bundesministerium für Ernährung, Landwirtschaft und Forsten (BMELF). Schriftenreihe des BMELF, A, Angewandte Wissenschaft, 452, Bonn 1996.
6. Zur Neuorientierung der Landnutzung in Deutschland. Opinion of Wissenschaftlicher Beirat beim Bundesministerium für Ernährung, Landwirtschaft und Forsten (BMELF). Schriftenreihe des BMELF, A, Angewandte Wissenschaft, 453, Bonn 1996.
7. Die Entwicklung der Landwirtschaft in Mitteleuropa und mögliche Folgen für die Agrarpolitik in der EU. Opinion of Wissenschaftlicher Beirat beim Bundesministerium für Ernährung, Landwirtschaft und Forsten (BMELF). Schriftenreihe des BMELF, A, Angewandte Wissenschaft, 458, Bonn 1997.
8. Zur Weiterentwicklung der EU-Agrarreform - Entkopplung der Preisausgleichszahlungen und Umsetzung der GATT-Beschlüsse. Opinion of Wissenschaftlicher Beirat beim Bundesministerium für Ernährung, Landwirtschaft und Forsten (BMELF). Schriftenreihe des BMELF, A, Angewandte Wissenschaft, 459, Bonn 1997.
9. Kompetenzverteilung für die Agrarpolitik in der EU Opinion of Wissenschaftlicher Beirat beim Bundesministerium für Ernährung, Landwirtschaft und Forsten (BMELF). Schriftenreihe des BMELF, A, Angewandte Wissenschaft, 469, Bonn 1998.
10. Integration der Landwirtschaft der Europäischen Union in die Weltagrarwirtschaft. Opinion of Wissenschaftlicher Beirat beim Bundesministerium für Ernährung, Landwirtschaft und Forsten (BMELF). Schriftenreihe des BMELF, A: Angewandte Wissenschaft, 476, Bonn 1998.
11. Zur Wettbewerbsfähigkeit der deutschen Milchwirtschaft. Opinion of Wissenschaftlicher Beirat beim Bundesministerium für Ernährung, Landwirtschaft und Forsten (BMELF). Schriftenreihe des BMELF, A: Angewandte Wissenschaft, 486, Bonn 2000.
12. Finanzpolitische Reformmaßnahmen des Bundes 1999 bis 2002. Zur Wettbewerbsfähigkeit der deutschen Milchwirtschaft, 488, Bonn 2000.
13. Beschlüsse des Rates der Europäischen Union zur Reform der gemeinsamen Agrarpolitik vom 26. Juni 2003. Opinion of Wissenschaftlicher Beirat beim Bundesministerium für Ernährung, Landwirtschaft und Forsten (BMELF). In: Berichte über Landwirtschaft, 2, 165-172.
14. Nutzung von Biomasse zur Energiegewinnung – Empfehlungen an die Politik. Opinion of Wissenschaftlicher Beirat beim Bundesministerium für Ernährung, Landwirtschaft und Forsten (BMELF). In: Berichte über Landwirtschaft, 216, Sonderheft, 2008.

15. Mitteilung der Kommission an den Rat und das Europäische Parlament zur Vorbereitung auf den GAP- Gesundheitscheck. Opinion of Wissenschaftlicher Beirat für Agrarpolitik beim Bundesministerium für Ernährung, Landwirtschaft und Verbraucherschutz (BMELV). In: Berichte über Landwirtschaft, 86 (1), 2008, 5-10.

16. EU-Agrarpolitik nach 2013. Opinion of Wissenschaftlicher Beirat für Agrarpolitik beim Bundesministerium für Ernährung, Landwirtschaft und Verbraucherschutz (BMELV). Mai 2010.

www.ingramcontent.com/pod-product-compliance
Ingram Content Group UK Ltd.
Pitfield, Milton Keynes, MK11 3LW, UK
UKHW021127160426
5217IPUK00046B/43